What are the odds?

From being hit by
a space rock to buying
a little happiness

Surendra Verma

ISBN: 9781511785037

Copyright © Surendra Verma 2015–2021

First published 2015
Revised edition published 2017
This revised edition published 2021

May the odds described in this book be ever in your favor.
(with apologies to Suzanne Collins, *The Hunger Games*)

Contents

Introduction

Applying reason to randomness

I suspect the truth is that we are waiting, all of us, against
insurmountable odds, for something extraordinary to happen to us.
– Khaled Hosseini, And the Mountains Echoed

In the 17th century, two games of dice popular were popular in the
casinos of Europe. In the first game, roll a dice four times and win
if a six would come; in the second game, roll two dice 24 times and
win if double six would come. Which game you would put your
shirt on?

The Chevalier de Méré – a high-living French nobleman and
gambler described by his friend the mathematician Blaise Pascal as
a man having a very good mind but no mathematics – thought the
chances of winning would be better in the second game. But when
the keen gambler continued losing his money, he asked his friend:
Why? Pascal worked with fellow mathematician Pierre de Fermat
to apply reason to randomness. Their hard analysis gave birth to
the theory of probability.

The theory which originated in a gamblers' dispute is now at
the heart of many enterprises which are more important than
gambling, including all kinds of insurance, educational
measurements and much of modern physics. (In quantum physics,
we can only calculate probabilities, while in pre-quantum classical
physics we calculated what happens. Einstein hated the

probabilistic nature of quantum physics. He thought that the universe was deterministic in some ways. He was fond of saying, "God does not play dice.")

Chance is something that happens in an unpredictable way. Probability deals with the chances of an event happening in an unpredictable way. By snatching numbers from fortune tellers and giving them to mathematicians it helps us to cope with uncertainty. (In this book the terms odds, chances, likelihood, and probability are used synonymously – *see* next page.)

This book describes odds of numerous events as interesting stories that are short, sharp, and simple – and, above all, relevant to everyday life. Stories are not only packed with up-to-date and thoroughly researched information, but they also present fascinating insights into the topics discussed.

In this book, you'll not find endless lists of trivial and meaningless odds or ways of calculating them. Although irresponsible at times, the internet can do this job well if you know how to separate the wheat from the chaff. If you Google "What are the odds," you would instantly be gratified with a long list of websites, most of them touting information that is either false or fossilized (their rightful place is in a garbage bin or a museum of antiquities). Mr. Google would also direct you to Professor Wikipedia bursting with information written in wiki-speak that requires a PhD to interpret. No such pitfalls when you're reading this book.

Happy reading the odd odds of our lives!

Note

Terms and units used in this book

In this book the terms odds, chances, likelihood, and probability are used synonymously. Simply put, the probability is the ratio of favorable outcomes to total outcomes, and it's written as a number between 0 and 1. A probability of 0 means there is absolutely no chance that event happening, a probability of 0.5 means there is a 50% chance and a probability of 1 means there is a 100% or absolute chance.

Traditionally, the term "odds" refers to the ratio of favorable outcomes to total outcomes. Bookmakers' way of writing odds is opposite to how we express the likelihood of winning with probabilities. Bookmakers' odds of 3 *to* 1 (also written as 3:1 or 3-1) means that there are 3 chances of losing and only 1 chance of winning, which means chances of winning are 1 *in* 4 (probability 0.25).

The term "1 in 4" (probability 0.25 or 25%) is easier to understand, and this is the way odds are expressed in this book.

The table below shows some of the terms commonly used to describe the probability of an event.

Term	Probability expressed as percentage	Probability expressed as odds
Absolutely certain	100%	1 in 1
Virtually certain	99%	95 in 100
Very likely	90%	9 in 10
Quite likely	70%	7 in 10
Evens (equally likely)	50%	1 in 2
unlikely	30%	3 in 10
Not very likely	20%	1 in 5
Extremely unlikely	5%	1 in 20
Never (absolutely no chance)	0%	0

Also, in this book the so-called short-scale values are used for:

1 billion = a thousand million = 1,000,000,000
1 trillion = a thousand billion = 1,000,000,000,000

Appendix at the back of the book describes some very simple mathematics of calculating probabilities. You don't have to read it to understand this book.

1

Death from the sky

**Odds of you being pulverized by a space rock are about 1
in 100 million; you're more likely to die in some other
natural catastrophe such as an earthquake, flood, storm,
or lightning. The odds of a giant asteroid (10-km or 6.2
mi wide) hitting the planet in the next 100 years and
wreaking unimaginable havoc are 1 in one million.**

Asteroids are giant rocks that orbit within a vast doughnut-shaped
ring between Mars and Jupiter. About a million are 1 km or more
in diameter, and billions are of boulder or pebble size.
Occasionally a collision kicks an asteroid out of its orbit, sending it
towards the Sun. These stray asteroids take up an orbit that loops
past the Earth. There may be as many as 100 million stray
asteroids larger than 20 m. One of these asteroids could destroy a
large city.

A few years ago, an errant asteroid 30-m (98-ft) across came
within 27,700 km (17,200 mi) of the Earth. If it struck the Earth, it
would have caused a cataclysmic explosion. In 1908, an asteroid of
similar size exploded in the Siberian sky. A thousand times the
force of the Hiroshima bomb, it flattened an area of remote
Tunguska Forest bigger than metropolitan New York and caused
the Trans-Siberian Express to shake widely on its tracks 600 km
(372 mi) from the explosion site.

Collisions of this magnitude occur once or twice every 1000

years. No person in the past 1000 years has been directly killed by a space rock though there are ancient Chinese records of such deaths.

Any asteroid over 3 km (1.8 mi) across is considered a global catastrophic risk. The odds of such a big space rock striking this century are about 1 in 200,000.

A giant asteroid did wipe out the dinosaurs, but that was 65 million years ago. This asteroid, 10 km (6.2 mi) across, gouged a crater 80 km (49 mi) wide and 30 km (18.6 mi) deep – big enough to devour the largest city. The odds of such an impact are 1 in 1 million in the next century, but an impact of this magnitude could temporarily destroy civilization.

Sleep easy, NASA assures us that at present no asteroid or comet is on collision course with the Earth.

2

Left out in a right-handed world

The odds of being left-handed are about 1 in 10 or 10%. This appears to be universally true for all human populations anywhere in the world. More men are left-handed than women. Very, very few people are truly ambidextrous.

Well, with such odds it's quite clear that most of the global population is right-handed. Even the Bible reminds us that this world is designed for right-handed (and righteous) people: On Judgment Day, the sheep (good people) at Jesus' right hand will be blessed by God while the goats (sinners) at his left will be condemned to eternal damnation (Matthew 25:31-46).

Most right-handed people are right sided overall: they mostly prefer to use their right foot and eye. A little over half are also right eared.

Why are more people right-handed? Writing in *Scientific American*, M. K. Holder, a US scientist, says there is evidence of for genetic influence, but geneticists cannot agree on the exact process. There is also evidence that handedness can be influenced by – and changed – by social and cultural factors. "For instance, teachers have been known to force children to switch from using their left hand to using their right hand for writing," she says. "Also, more restrictive societies show less left-handedness in their populations than other more permissive societies."

There is no hard evidence of any non-human species showing species-level handedness found in humans.

3

So, you want to be a millionaire (without lifting a finger)

The odds of picking up six winning numbers out of 45 in a lottery are 1 in 8 million (8,145,060 to be exact); the odds increase to 45,379,620 if you must pick up seven winning numbers.

The odds of selecting a ball with a certain number, say 12, at random from a box containing 45 balls numbered 1 to 45, is 1/45. If you must randomly select six balls, the odds increase to 1 in about 8 million. The odds increase to 1 in about 14 million if you must randomly pick six balls from a box containing 49 numbered balls, and to 1 in 19 million if the box has 51 balls. Even though the odds are stacked against them, millions of people buy lottery tickets every day. Sure, to win you must be in it.

The Roman Emperor Augustus probably conducted the first public lottery for a community cause, raising funds for repair work in the city of Rome. These lotteries were simple as they involved drawing a number from a few hundred or a few thousand. Lotteries these days are complex affairs. Here're some of the myths of modern lotteries.

The larger the jackpot, the poorer the odds of winning: The odds of winning a lottery never change. But it is possible that because of a large jackpot, more people will buy tickets. This will increase the odds of sharing the jackpot.

There are lucky numbers and unlucky numbers: Every number has the same chance of being drawn. Even "strange" combinations such as 1, 2, 3, 4, 5, 6 have the same chances of winning as any other combination, say, 5, 12, 29, 31, 38, 40. Some newspapers publish lists and charts of the frequency of numbers drawn. But, according to the theory of probability, these calculations are meaningless. It is a waste of time to use past numbers to predict future ones. Similarly, there are no "hot" (a number that come up in the past few draws) or "due" (a number that has not appeared for a while) numbers.

Lottery systems can help you pick the winning numbers: There are no lottery systems or winning strategies that can help you win the jackpot. If there were such systems, all mathematicians would be very wealthy people. They would not be teaching in schools or colleges. Do not waste time on "killer strategies" that promise 'to give you an edge over people who just randomly pick numbers or even worse buy computer-generated quick picks. No computer, no fancy formula, no book can help you change the odds of winning a lottery.

You can improve your odds of winning in a particular lottery by buying more tickets: In fact, more tickets you buy more money you lose.

If you do really want to be a millionaire, bet on the odds-on Australian favorite *hard yakka* (hard work).

4

Winning numbers

Can entropy help you boost your chances of winning the jackpot in a national lottery?

In national lotteries where you must six numbers out of 45 or 49, most players do not select numbers at random; they usually go for their favorite or lucky numbers. There is also a psychological tendency for players to spread their choices. Would your odds of winning change if you always chose the numbers 1, 2, 3, 4, 5 and 6?

The chances of winning do not change whether you select the numbers 1, 2, 3, 4, 5 and 6 or any set of six random numbers. A few years ago, boffins at the University of Southampton in the UK used a supercomputer to analyze the UK national lottery in which players pick six numbers out of 49. They advised players to avoid the sequence 1, 2, 3, 4, 5 and 6 because 10,000 people a week select this combination. The chances of winning do not change, but if the numbers 1, 2, 3, 4, 5 and 6 do come up, the winner would have to share the prize money with all those people who are either lazy or correctly think that it is meaningless to impose "lucky" numbers on random numbers.

The University of Southampton researchers used entropy, a concept normally applied to many atoms, to their analysis of the numbers chosen by lottery prize winners to assess how players make their selections. Entropy is a measure of disorder or randomness of a system. The more random or disordered a system is, greater the entropy. The "maximum entropy" technique the researchers used helped them to deduce information buried in data.

The University of Southampton study and other studies reveal that your choice of numbers will not affect the odds of winning but the following will maximize any win.

- Picking numbers on the edges and the lower part of the game card will usually increase the size of any prizes that are won, as people tend to shy away from these.
- Choosing a pair of consecutive numbers, such as 31 and 32, is a good strategy as people tend to spread their lottery number out.
- Choosing pairs of numbers greater than 31, such as 41 with 44.
- Avoid choosing numbers below 31 as these are frequently picked to match birthdays.
- Restrict your choices to prime numbers greater than 10 (that is, numbers 11, 13, 17, 19, 23, 29, 31, 37, 41, 43 and 47).

Here's our disclaimer in fine print:

There is no scientific way to predict winning number in advance, and, in general, buying lottery is not an "investment."

5

Are we alone?

As astronomers continue to discover more Earth-like planets the odds of finding a cosmic neighbor are fast decreasing, but don't bet your house on it.

In recent years, astronomers have discovered more than 1,800 planets beyond the solar system (estimates suggest that our galaxy alone has at least 100 billion more). Of these 1,800 exoplanets, about 20 are Earth-like and therefore capable of hosting life. This leads to the tantalizing question: Is there extra-terrestrial intelligent life? The following is one of the ways of calculating the odds.

Some astronomers say that the probability of a star having a planetary system is close to 1, but we can assume this figure to be 1/2. The probability of a planet being Earth-like is 20/1,800 or 1/90. DNA is an extremely complex molecule with a very small chance of occurring on its own and that life is precarious because the universe is a dangerous place. By this reasoning, we can assume the probability of life occurring on any single planet that is Earth-like is extremely remote and can assigns an arbitrary value of 1 in a trillion. By multiplying 1/2, 1/90 and 1/trillion, we arrive at a figure of roughly 0.000,000,000,000,000,5 which is the probability of life around any one given star.

There are probably1,000 billion billion stars in the universe, and if put this value in the rule for combining the probabilities of independent events, the probability of life on at least one other planet outside Earth would be $(1 - 0.000,000,000,000,000,5)$ multiplied to itself 1,000 billion billion times and then the calculated value subtracted from 1.

This number is very close to 1. The odds point to virtual certainty that there is life on planets beyond Earth.

6

If a monkey wrote this book

What are the odds of six monkeys strumming unintelligently on keyboards for millions and millions of years writing this book?

If you're keen to work out the mathematical probabilities of a monkey writing this book or for that matter any book, Google "infinite monkey theorem."

Assuming Shakespeare's Hamlet contains 130,000 letters (plus punctuation marks and capitalization), the odds are one in 4 x 10360,783 (4 followed by 360,783 zeroes) to get the text right at first trial (from Wikipedia, after translating from wiki-speak). Even if the observable universe were filled with monkeys the size of atoms typing from now until the end of the universe, they have many magnitude* less than one chance in 4 x 10360,783 of writing Hamlet.

This answer is beyond the comprehension of any monkey – and the author.

In 1930, the English scientist and mathematician James Jeans wrote in *The Mysterious Universe*: "Six monkeys, set to strum unintelligently on typewriters for millions and millions of years, would be bound in time to write all the books in the British Museum."

He continued to say, "... but if we looked through all the

millions of pages the monkeys had turned off in untold millions of years, we might be sure of finding a Shakespeare sonnet somewhere amongst them, the product of the blind play of chance. In the same way, millions of millions of stars wandering blindly through space for millions of millions of years are bound to meet with every sort of accident, and so are bound to produce a certain limited number of planetary systems in time."

Life on Earth is also "the product of the blind play of chance" and Jeans' six monkeys are also capable of writing the sonnet of life on other planetary systems. As we said if in the previous story, we are not alone.

---

* An approximation to the nearest power of 10 of a number is called its order of magnitude. For example, a dollar (= 100 c = 102 c) is two order of magnitude more valuable than one cent.

7

Bridge is an odd card game

Your chances of being dealt a bridge hand containing 13 cards all the same suit (in any order) are 1 in 158,753,389,900.

If you are a bridge player, you have probably heard this joke.

> A cleaning lady applying for a new job was asked why she had left her last one.
> She replied, "They paid me good money, but it was the strangest place I ever worked. They played a game called bridge. Last night there were a lot of folks there and overheard a woman say to another woman, 'You play with my husband, and I'll play with yours."
> Then I heard a man say to a woman, "Take hands off my trick." I nearly dropped dead when a woman said, "You forced me, you jumped twice and then you didn't have the strength for one raise."
> The man replied that he had the strength but not the length and woman said she was protecting her honor.
> As I was leaving, I heard a woman say, "I guess we have to go now – this is the last rubber."

The joke is a play on bridge terms: trick (four cards, one from each player), jump (a bid that is one level higher than necessary), strength (cards with more points), length (number of cards held in a suit), raise (to bid for more tricks), honor cards (aces, kings, queens, jacks, and tens) and rubber (a unit of scoring).

Experts disagree on how any times a pack of 52 cards be shuffled before the cards are randomly distributed. Their answers lie between five and seven times. Three lazy shuffles are definitely not enough. Once you have shuffled the pack thoroughly, you're ready to learn more odds for bridge hands.

- Four aces: 1 in 377
- No honor card: 1 in 1,827
- 13 spades in any order: 1 in 635,013,559,600
- 13 spades in correct order (AKQJT98765432): 1 in 4 billion trillion
- All four players getting one full suit each: 1 in 2,235 trillion trillion (*see* next page)
- All four players getting all 13 cards of their favorite suit in the correct order: the odds are astronomical with enough naught to fill a couple of lines of this book

If your pack of 52 cards has been shuffled so well that the pack is truly randomized, the odds of the cards ending in perfect order spades from K to 2, then hearts, diamonds and clubs in the next shuffle are around 1 in 1068 (1 followed by 68 zeroes).

8

Whimsical whist

**The odds of all players in a whist game being dealt a
perfect hand (each containing a full suite of 13 cards) are
1 in 2,235 trillion.**

Or exactly 1 in 2,235,197,406,895,366,368,301,599,999. If the
entire population of the world played whist all day long for several
lifetimes, the odds against this happening would still be equivalent
to several lifetimes.

In January 1998, a group of whist players at a whist club in
Suffolk, UK were gob smacked when each of them received full
suit in one hand. And it happened again in November 2011 to
another group of whist-playing pensioners in Warwickshire, UK.
It's like finding a specific drop of water in the Pacific Ocean –
twice. Unbelievable!

Whist is a classic English card game which dates to the 17th
century. It's played by four and the players sitting across from each
other are a team. Each player is dealt 13 cards and the teams
compete to win tricks (four cards, one from each player, is one
trick). Unlike bridge, the rules are extremely simple but there is
enormous scope for scientific play.

Two bob each way on a game of two-up

Our two bob's worth on the iconic Australian gambling game.

Once expressing my opinion on a German scientist's theory, I wrote something like "here is my two bob's worth" on the theory. The scientist sent me an email asking the meaning of "bob." I replied that in the pre-decimal currency days in Australia "bob" was an informal word for a shilling (made mostly of silver), and the phrase "two bob's worth" (which has now turned into "wo cent's worth") means opinion or say. He wasn't very happy, not with my opinion on his theory, but the value of such a low sum I placed on my opinion. No, my opinions are not "two diamond's worth." Perhaps meaning was lost in translation.

Similarly, those who are not familiar with Australian history would fail to understand what the game of two-up meant to Australian soldiers who played it extensively during World War I. In honor of the diggers (a term of endearment used for Australian soldiers, especially ones who served in World War I), it is not unlawful in many states of Australia to play the game on Anzac Day which is celebrated on 25 April every year.

First, to "have two bob each way" means to support two contradictory opinions or causes at the same time, often in self-protection. No point in betting head and tail at the same time in a game of two-up.

In two-up, two coins (traditionally old pennies as they are much heavier than new coins and spin well) are spun in the air and bets are laid on whether the coin faces show heads or tails. The spinner (a person selected to spin the coin) places two coins on a kip (a small wooden bat) and tosses them in the air. The spinner is required to place a bet before their first throw and the bet must the equaled by another player. The results are:

- Two heads: the spinner wins – probability 25%
- Two Tails: the spinner loses – probability 25%
- Odds (one head, one tail): spinner throws again – probability 50%
- Odding out (spinning five "odds" in a row) – 3.125%

Although coin tossing is a random event, but naturally tossed coins also obey the laws of mechanics and their flight is determined by their initial conditions. This could affect true probabilities shown above, although in a very minor way.

Coincidences: the odds of happening them

If you know the laws of probability, you wouldn't be enchanted by coincidences.

The name of a friend who you haven't seen for years suddenly pops up in your mind, you turn your computer on and there is an email from her. In your dream, you see a lottery ticket and the next day you win a prize in a lottery. We all have had a premonition or a dream that later became true.

The Swiss psychologist Carl Jung coined the term "synchronicity" to describe the simultaneous occurrence of events that were linked but couldn't be explained by the law of cause and effect. He said that such coincidences were meaningfully related; they were not random events happening in accordance with the laws of probability.

Jung was a colleague of Sigmund Freud but broke with him to work on his own psychological theory. To find evidence for his idea of synchronicity, Jung analyzed horoscopes of four hundred married couples. He could not find any significant statistical support for the existence of a connection without any cause between horoscopes and actual marriages, yet he clung to the idea. His idea of meaningful coincidences made some sense in his nineteenth-century world, a world saturated with spiritualism and the occult. But not now.

We all are fascinated by coincidences, but "meaningful coincidences" are not remarkable but random coincidences. The following coincidences seem weird, but they are true.

Abraham Lincoln was elected to US Congress in 1846.
John F. Kennedy was elected to US Congress in 1946.

Lincoln was elected US President in 1860.
Kennedy was elected US President in 1960.

Andrew Johnson, who succeed Lincoln, was born in 1808.
Lyndon Johnson, who succeeded Kennedy, was born in 1908.

Google "Lincoln + Kennedy" and you will find a few more coincidences linking them. To understand the odds of such random coincidences occurring, let us the take the classic birthday problem: how many people should there be in a room for it to be more likely than not that at least two of them share a birthday, say 26 January. There are 365 days in a year, or 366 if it's a leap year. If there are 367 people in the room, a coincidence is guaranteed. How many people are needed to ensure 50% chance of a shared birthday? You're likely to say half of 366 or 183. But the correct answer is 23. For a 95% chance of match, we need 45 people.

Even in a small group of 45 people there is a 95% chance of at least one birthday overlap. Next time you find yourself in a room with 45 people, you can almost safely bet a few dollars that at least two people in the room have the same birthday.

The law of truly large numbers says that anything remotely possible will eventually happen, if we wait long enough. "Before you interpret your hole-in-one as a sign of special favor from the Gods of Golf," says British mathematician Ian Stewart, "do bear in mind that there are an awful lot of golfers, and according to the laws of probability such an event will, every so often, happen to one of them." The odds of getting a hole in one are 1 in 5,000, as calculated by an unknown mathematician.

Bookmakers' odds

Bookmakers' jargon for occasional punters.

Bookmakers' way of writing odds is opposite to how we express the likelihood of winning with probabilities. Bookmakers' odds of 3 to 1 (also written as 3:1 or 3-1) means that there are 3 chances of losing and only 1 chance of winning, which means chances of winning are 1 in 4 (this is the way odds are expressed in this book). Here're some of the terms used by bookmakers.

- *Evens*: Odds of 1 to 1. The odds of winning or losing are exactly equal.
- *Odds against:* Odds longer than evens. Odds of 3 to 1 mean that when you bet $1 the bookmaker would pay you $3 plus your outlay of $1 (that is, a total of $4) if you win.
- *Odds on*: Odds shorter than evens; odds are such that the profit of the bet being less than the original bet. Placing a winning bet of $1 on odds of 1 to 2 would give you a return of 50 cents plus the outlay of $1, that is, $1.50.
- *Long odds*: Odds such as 100 to 1 offered against a competitor unlikely to win.
- *Short odds*: Odds such as 3 to 2 offered against a competitor who has a good chance of winning.
- *Favorite*: The competitor with the shortest or lowest odds

because it's likely to win.

You might have to jump off the bridge if you bet on races and don't understand these horse racing terms.

- *Win*: Your horse must finish first to collect.
- *Place*: Your horse must finish first or second to collect.
- *Quinella*: You bet on two horses in a race which must finish first and second in either order
- *Exacta*: You bet on two horses in a race which must finish first and second in exact order.
- *Trifecta*. You bet on three horses in a race which must finish first, second and third in exact order.
- *Superfecta*: You bet on four horses in a race which must finish first, second, third and fourth in exact order.
- *Daily double*: A bet that requires selecting the winners of two nominated races at a race meeting.
- *Quadrella*: A bet that requires selecting the winners of four nominated races at a race meeting.
- In *fixed odds* betting, the price of the horse when you place the bet is the price you get.
- In *parimutuel* betting, prices change as the amount of money on each horse is invested; the more money on a horse, shorter the odds will be.
- A *"jump off the bridge"* wager is so high that you feel like jumping off the nearest bridge if your horse loses.

Bookmakers' odds are always determined in such a way that only the bookmaker can win during the continuing conduct of the game. The odds posted by a bookmaker are not "true" odds, but the payout odds. Let's see how the system works. Suppose the true odds in a match look like this:

Team Red to win	Odds 5/3	37.5% chance
Team Blue to win	Odds 5/3	37.5% chance
Draw	Odds 3/1	25% chance

As the above probabilities add to 100%, bookmakers will skew the odds in their favor and post the following payout odds.

Team Red to win	Odds 6/4	40 % chance
Team Blue to win	Odds 6/4	40 % chance
Draw	Odds 12/5	29.4 % chance

The new odds overestimate the chance to win or draw to ensure that the punter is always underpaid, but they ensure that bookmakers make 9.4% profit.

You know now how the betting works before you put $100 each way (to come first, second or third) on "Punter's Misfortune" in the Kentucky Darby or the Melbourne Cup.

Playing for spiritual even odds

It's a different kind of wager.

Blaise Pascal, the brilliant 17th-century French mathematician, physicist, and philosopher, is known for many discoveries in physics and mathematics, including the probability theory. But he is now more widely known for *Pensées* (Thoughts), a collection of meditations on the nature of human life.

He makes at least one application of the theory of probability in one of his meditations, now known as Pascal's Wager. God either is or He is not. If we bet on whether God's exists, there are two chances. If we win the bet the reward is enormous (eternal happiness) and the loss is insignificant (only the time we must spend in worship). If we lose the bet, still we lose little. Believing in God is the more sensible wager.

If the agnostic remains unconvinced, Pascal offers a different way of looking at the odds. If there were infinite eternal happiness to be won by leading religious life, there is one chance of winning against a finite number of chances of losing. Your chances of winning or losing are finite. You are playing for even odds as there are as many chances on one side as on the other. Therefore, it will pay to lead a religious life.

Pascal said, "the heart has reason, which reason cannot know." Therefore, our belief in God cannot be explained by science's reasoning.

But we must still follow the course of reason, reason which is reasonable and respects creativity and morality.

Our clever cousins, Neanderthals

If you're of non-African origin, you're an odds-on favorite to share about 2% of your DNA with Neanderthals.

Neanderthals suffer from bad press. Their name is synonyms with boorishness, brutishness, idiocy, ill-manners, or plain stupidity. The origin of this image in pop culture can be traced back to 1911 when the first Neanderthal skeleton found in France appeared to have a curved spine, a stoop, bent knees and a head and hips that jutted forward. Re-examination of the skeleton in 1957 showed that the original owner had suffered from a grossly deforming type of osteoporosis; the skeleton didn't represent the average Neanderthal.

We know now that Neanderthals were not brute cavemen but had minds like our own capable of abstract thinking. The blood lines of Neanderthals and *Homo sapiens* mingled when they had completed their long migration from Africa into the Eurasian continent. Both species lived in Eurasia between 350,000 and 39,000 years ago.

Recent studies reveal that 1.5 to 2.1% of DNA of non-Africans today comes from Neanderthals. DNA is the inedible genetic code which tells us who we are and where we come from. Any given non-African carries a small amount of Neanderthal DNA, but not everyone carries the same bits. The flip side of these bits of Neanderthal DNA is that it makes the carriers susceptible to type 2 diabetes, Crohn's disease, and addiction to smoking.

Neanderthals disappeared around 39,000 years ago. The cause of their extinction is not well understood. Did they fail to adapt to the last Ice Age, or did they become absorbed in our gene pool? Neanderthals were our sister species; their demise allowed *Homo sapiens* to flourish.

Smart, smarter, smartest

The odds you are having an IQ of 202 or higher are 1 in 200 billion. Don't despair: the odds of you joining Mensa, the international high IQ society, are merely 1 in 50.

Real-world intelligence has many dimensions. We can be more intelligent in some things and not others. No one is equally intelligent in everything. There's no absolute measure of all characteristics of intelligence. The most famous – not necessarily the perfect – measure is the IQ tests.

In case you don't know, IQ or intelligence quotient is the ratio of actual mental age, as measured by intelligence tests, to the mental age that is normal for a particular chronological age. The average score is 100, and 70% of people have an IQ between 85 and 115. Genius range starts from around 135; only about 1% of all the people in the world have an IQ higher than 135. Einstein is believed to have an IQ of 160.

Intelligence is not related to the size of the brain or the numbers of neurons in the brain. Some anatomical and physiological differences occur in the male and female brain, but men and women show no consistent difference on IQ tests.

Mensa members have a score at or above the 98th percentile on a standard IQ test (a score that is greater than that achieved by 98% of the general population taking the test). If you are pining for the membership of Mensa, you would be pleased to know that a growing number of studies show that mental and physical exercises can help raise your IQ score by a few points.

15

Born in blood

Hemophilia, a rare bleeding disorder in which the blood doesn't clot normally, affects males more frequently (1 in 10,000) than females (1 in 100,000,000).

Hemophilia is usually passed from parents to children through genes. It can also be acquired, meaning people are not born with the disorder but it develops during their lifetime. People are born with hemophilia; they cannot catch it from others.

The blood clotting genes for hemophilia are carried on the X chromosome. Males carry XY chromosomes, females XX. A child inherits one X chromosome from the mother and either the X or Y chromosome from the father. An X makes the child female, Y a male. Thus, there are 1 in 2 chances (50%) of being a born as a boy or a girl.

Since males carry only one X chromosome and if this chromosome is carrying hemophilia, it will immediately show up. If a woman has one hemophilia-carrying chromosome, she will still have normal blood clotting, but she will be a carrier of the hemophilia gene. If father doesn't have hemophilia but mother is a carrier of the hemophilia gene, there is a 50% chance that sons will have hemophilia, and 50% chance daughter will carry the hemophilia gene. If mother is not a carrier of the hemophilia gene but father has hemophilia, there is 100% chance daughters will be a carrier of the hemophilia gene, and absolutely no chance sons will inherit hemophilia.

Hemophilia is sometimes described as the "royal disease" because of its widespread occurrence in the royal families of Europe. Queen Victoria was a carrier of hemophilia. She passed it on to one son, Leopold, who died of bleeding, and two daughters, Alice and Beatrice. Her other three sons, Edward, Alfred, and Arthur were unaffected. Alice passed it on to her daughter Alexandria who married Nicholas II, the last tsar of Russia. The 'mad monk' Rasputin claimed he could treat their son Alexei's hemophilia. Rasputin was somehow able to relieve Alexei's sufferings. As a result, Rasputin received an unlimited trust from the tsar and his wife. Rasputin's great influence on them played a significant part in the tsar's unpopularity which contributed to the Russian revolution. Alexei didn't die of hemophilia; he was executed at the age of 13 with the rest of his family by the revolutionaries.

In 2009 DNA tests on Alexei's remains showed that Queen Victoria carried hemophilia B. The other type is A. The difference in the two types is in the clotting factors which are specialized proteins. About 8 out of 10 people who have hemophilia have type A.

Cancer: everywhere but not out of control

The odds of one day anyone of us being diagnosed with some forms of cancer are 4 in 10; in other words, 40% of us will develop cancer at some point in our lives. Healthy living is not a guarantee against cancer, but it certainly stacks the odds in your favor.

Cancer, "the emperor of all maladies," as Siddhartha Mukherjee describes it in his Pulitzer-winning book of the same name, is caused by damage to our DNA which stores genetic information and passes it on to the next generation by making an identical copy of itself. Cancer can be inherited (tumor-causing genes handed down from our parents) or environmental (the sun's UV rays, tobacco or some other chemicals that damage our DNA).

The prestigious journal *Science* recently published a landmark study that said that 65% of the variations in cancer rates in different tissues are the product of random mutations accumulating in healthy stem cells. Most media outlets, including *Time* magazine, interpreted this study as showing that 65% of all cancers are the results of pure chance or sheer bad luck. It's not true. Cancer is not out of control.

More than 30% of all cancers – there are more than 100 types – worldwide can be prevented by lifestyle changes such as not smoking, keeping a healthy body weight, keeping active and staying safe in the sun, and immunization against cancer causing infections, hepatitis B virus (HBV) and human papillomavirus (HPV). Many cancer types can be cured if detected early and treated adequately.

Tobacco use is the single largest preventable cause of cancer in the world causing 22% of cancer deaths.

In 1975, the lifetime risk of being diagnosed with cancer was about one in four people or 25%. This figure has now risen to 40%. Overall cancer rates are higher among men than women. In 1975, the lifetime risk to women was slightly lower than men. In 2030, the lifetime risk for women is projected to be 44% for women and 50% for men.

In the 20th century, about 530 million people died of cancer. Compare this figure with 1,970 million deaths caused by other non-communicable diseases (of which 1,246 million cardiovascular diseases); 1,680 million by infectious diseases; and 980 million by accidents, murder, wars, drugs, and air pollution.

Your misadventures measured in lost microlives

A better way to calculate the odds of consequences of your good and bad habits.

We all know how lifestyle choices such as smoking, drinking, eating junk food and not exercising at all badly affect our lives, but how to compare their risks? To answer this question, David Spiegelhalter, a renowned British risk-management expert, has come up with a simple but unique unit of measurement: a microlife, which is 30 minutes of our life expectancy. He defines a microlife as the result of chronic risk that reduces life, on average, by just one of the million half hours we have left.

Data in the table below is based on lifelong habits of men and women, ages 35 and up, averaged over large populations. Over time healthy habits slow down ageing; unhealthy habits accelerate ageing. "No one likes to get older faster," says Spiegelhalter. Absolutely.

Microlives gained over one day		
Exercise (first 20 minutes)	2	Total microlives gained: 5 (Odds are in your favor)
Exercise (subsequent 40 minutes)	1	
Alcoholic drink (first 10 grams, or 0.35 oz, of alcohol)	1	

Fruit and vegetables (1.25 servings)	1	

Microlives lost over one day		
Overweight (each day, for each five kilograms above the optimum)	1	Total microlives lost: 6.5 (Odds are against you)
Sitting (two hours without activity)	1	
Smoking (six cigarettes)	3	
Red meat (80 grams, or 2.8 oz)	1	
Alcoholic drinks (each one beyond first)	0.5	

Spiegelhalter likes to divide risks that affect our lives into two groups: acute risks that could kill you, and chronic risks that accumulate trouble for future. The same risk might have both effects, he says, drink too much alcohol you might fall over and bang your head, or it might progressively damage the liver and send you to an early grave.

Micromort (from micro and mortality) is the unit used for measuring acute risk. Coined in 1979 by Ronald Howard of Stanford University, a micromort is one in a million chance of dying. A micromort reduces our lifetime expectancy by a millionth. A lifetime probability of dying is one mort, so one day costs about 39 micromorts for an average person.

For example, smoking 1.4 cigarette costs a micromort, the same as travelling 10 km (6 mi) by motorbike because of increased risk of an accident. But there is a difference. If you survive your motorbike ride, slate is wiped clean, and you start the next day with an empty account. Smoking, on the other hand, forces you to accumulate your microlives.

Micromort measures acute risk; microlife chronic risk.

It's now one hundred seconds to midnight

The odds are shortening on humanity's proximity to doom.

In 1947, the *Bulletin of the Atomic Scientists* introduced a symbolic clock to mark humanity's proximity to the apocalypse. It was set to seven minutes to midnight to evoke the imagery of apocalypse (midnight) and then popular image of nuclear explosion (countdown to zero). Closer to midnight we are, the worse off we are.

The Doomsday Clock, as it's known now, appears on the cover of every issue of the *Bulletin* and has become a universally recognized indicator of the world's vulnerability to catastrophe from nuclear weapons, climate change and emerging technologies in life sciences.

In December 2020, the clock's minute hand was moved from two minutes to midnight to 100 seconds to midnight – scarily close to a global meltdown. In January 2021, the *Bulletin* announced the Doomsday Clock remains 100 seconds to midnight. "The COVID-19 pandemic revealed just how unprepared and unwilling countries and the international system are to handle global emergencies."

Since its creation, the clock has been adjusted only 24 times. Its halcyon days were in 1971, just after the end of the Cold War, when it was sanguine 17 minutes away from midnight.

On the climate front, the odds of doomsday are certainly shortening if we look at these figures: Greenhouse gas emissions are now 50% higher than they were in 1990. Emission rates have risen since 2000 by more than in three decades combined. The year 2020, effectively tying 2016, was the hottest year ever measured, based on temperature records going back to 1880.

On the nuclear front, although Russia and the US no longer have the tens of thousands nuclear weapons they had during the Cold War, there is still a stockpile of some 16,000 nuclear warheads, 2,000 of them ready to launch in minutes. From 2009 to 2013, the US has cut only 309 warheads from its stockpile.

The memories of the Cold War are fading, but it's too much of wishful thinking that nuclear war wouldn't happen now. During a given year the odds of inadvertent nuclear war between Russia and the US are from 1 to 100 to 1 to 100,000, depending on various assumptions, according to Seth Baum of the Global Catastrophe Risk Institute, a non-profit think tank. An inadvertent nuclear war could happen when one side mistakenly believes it is under attack and launches what it believes to be a counterattack, but it is actually a first strike. The total annual probability for all types of nuclear war will be much larger than this, possibly much larger, Baum says.

19

Will you live to be a centenarian?

The odds that you will live to be 100 very much depend on geography (where you live) and sex (odds are dramatically higher for women than men).

Jeanne Calment, the Frenchwoman who died in 1997 at the age of 122, was probably the oldest living person ever. The chances of you and me living to her age are like our chances of winning ten eight gold medals in a single Olympics.

If we want to beat Madame Calment's record, choosing the right ancestors would be a big help. They should be rich (wealth brings health) and should have right genes (not the genes with a predisposition to diseases such as cancer or Alzheimer's).

In general, your chances of reaching 100 are better if you live in a developed country, and if you are a woman (women make up 82.8% of individuals age 100 years and older while men make up just 17.2%). Of course, a healthy lifestyle helps (Madame Calment credited her long life to her sense of humor, "I'll die laughing," she said).

Overall, life expectancy is increasing throughout the world because of advances in medical research. According to the World Bank's 2019 data, life expectancy in the world is 72. In UK it increases to 81, Australia to 83 years; but in the US the figure is slightly lower at 79 years.

In terms of centenarians (people aged 100 or over), Japan leads the world.

Country	Centenarians per million people
Japan	638
France	321
Italy	315
United States	304
United Kingdom	215
Germany	210
Australia	188
Russia	140
China	40
India	21
Entire world	**62**

20

Earthquake!

About 1,370 earthquakes happen around the world every year, but the chances of a great earthquake (magnitude 8.0 or greater) happening in any given year are 1 in 500,000. For mega earthquakes (magnitude 9.0 or greater) the odds increase to 1 in 11.4 million. Scientists agree that earthquakes of magnitude 10 or more are implausible.

G. K. Chesterton, sometimes described as the most unjustly neglected writer of the 20th century, advises that we should always endeavor to wonder at the permanent, not the mere exception. 'We should wonder less at the earthquake, and wonder more at the earth,' he says. But how can we not wonder at a mega earthquake – a mere exception – which causes awesome ripples in the ground which, in turn, cause immense destruction and untold sufferings for hundreds of thousands?

About 90% of the world's earthquakes occur in the 40,000 km (25,000 mi) long horseshoe shaped zone around the Pacific Ocean. Known as the Pacific Ring of Fire, it is home to about 75% of the world's active and dormant volcanoes.

The magnitude of an earthquake depends on the length of the fault (the crack in the earth's crust) on which it occurs – the longer the fault, the larger the earthquake. The famous San Andreas Fault, which runs along the west coast of the US, is about 1,300 km

(800 mi) long. To generate an earthquake of magnitude 10.5 would require the rupture of a fault many times longer than the San Andreas Fault.

The largest earthquake ever recorded was a magnitude 9.5 on 22 May 1960 in Chile; it was on a fault about 1,600 km (1,000 mi) long. The largest earthquake (magnitude 9.2) in North America happened on 28 March 1964 in Alaska. The biggest onshore Australian earthquake (magnitude 7.3) happened on 29 April 1941 at Meeberrie in Western Australia.

The Richter scale, ranging from 0 to 10, is used to measure the magnitude of an earthquake. The scale is logarithmic, which means each additional point represents a tenfold increase in the severity. Thus, a magnitude 5 earthquake is ten times as one of magnitude 4 and hundred times as powerful as on magnitude 3. (*See table*)

Magnitude on Richter scale	Intensity	Characteristic effects
less than 3.1	instrumental	detected only by seismographs
3.5	feeble	noticed only by some people at rest
4.2	slight	similar to vibrations caused by a heavy truck
4.5	moderate	people indoors feel movement; parked cars rock

4.8	rather strong	wakes sleeping persons; pictures on the wall move
5.4	strong	trees sway; furniture moves; some damage
6.1	very strong	walls crack; drivers feel their cars shaking
6.5	destructive	chimneys fall; weak structures damaged
6.9	ruinous	most houses collapse
7.3	disastrous	ground badly cracked; railway tracks bend
8.1	very disastrous	most buildings collapse; large landslides occur
more than 8.1	catastrophic	total destruction; ground moves in waves

21

The odds anti-vaxxers fail to appreciate

Not to drown you in streams of data, we present just one figure: the likelihood of more severe allergic reaction after MMR is 1 in 1 million, much less than being struck by lightning. Measles, the first letter in MMR (others stand for mumps and rubella), is so contagious that if one person has it, 90% of the people close to that person who are not immune will also become infected.

Before immunization of measles began it used to cause 2.6 million deaths a year worldwide. Now deaths from measles are as rare as hen's teeth, to use an old idiom. The highly publicized claims that MMR was associated with autism has been discredited by a slew of rigorous scientific studies. Surely, like any medical procedure, all vaccination, not only MMR, carries some risks as well as substantial, proven benefits.

Many parents perceive all types of vaccination as unsafe and unnecessary for their children. These anti-vaxxers present a threat to the success of vaccination programs. The World Health Organization dismisses these urban myths as FALSE.

- Better hygiene and sanitation will make diseases disappear – vaccines are not necessary.
- Vaccines have several damaging and long-term side-effects

that are not know. Vaccination can even be fatal.

- Some vaccines can cause sudden infant death syndrome.
- It is better to be immunized through disease than through vaccines.
- Giving several vaccines at the same time has adverse effects on children's immune system.
- Vaccine-preventable diseases are almost eradicated in my country, so there is no reason to be vaccinated

And it's false to believe that that influenza, commonly known as the flu, is just a nuisance and the vaccine isn't effective. According to the US Center for Disease Control and Prevention, flu vaccine reduces the odds of getting the flu by 70% to 90%. In older people, it is 80% effective in preventing death from flu complications. Immunity to influenza decreases over time and we need flu vaccine each year to continue to be protected.

The story of vaccination began in 1796 when Edward Jenner, an English doctor, took some fluid from the sore of a milkmaid suffering from cowpox and injected the fluid in the arms of a healthy eight-year-old boy (such experiment would be illegal today). Seven weeks later Jenner infected the boy with some fluid taken from the sore of a person suffering from smallpox. The boy did not show any symptoms of smallpox. Cowpox had given him immunity from smallpox. After tests on several other children, including his own 11-month-old son, Jenner published his findings in 1798. He called the process vaccination (from the Latin *vacca*, cow).

In the 18th century smallpox was the scourge feared by kings and peasants alike. Its victims, if survived, were left disfigured with severe scarring of the skin with deep pockmarks. Jenner had heard milkmaids saying that they cannot be infected by smallpox since they had had cowpox. Cowpox is a mild disease which causes minor sores on udders of cow, and anyone infected with it develops sores on their hands. Jenner decided to test whether there was any truth in this country wisdom.

But such down-to-earth wisdom is missing from the case against vaccination put forward by some parents. In the online version of *The Atlantic* magazine, Kimberly Christian-Campbell laments that we hear a lot about the risks of vaccination from anti-vaxxers but little about the vast majority of parents who quietly continue to vaccinate without complications. Deep rivers move with silent majesty; shallow streams make the most noise. To put it bluntly, down-to-earth wisdom encapsulated in this old proverb explains the noise made by anti-vaxxers.

Incidentally, in 1980 the World Health Organization declared the planet free of smallpox. A testimony to the success of a global immunization program. Anti-vaccine campaigners probably didn't exist in those days.

22

Calculating your chances of becoming a billionaire

You have about 1 in 3.5 million chance of becoming a billionaire worldwide; and only 1 in 21,500 of making the world's wealthiest 1%.

These odds are based on 2755 billionaires among the humanity's 7 billion or so members, according to business magazine Forbes' 2021 billionaire list which ranks individuals' assets. Of these, 724 live in the US, followed by China with 698 (including Hong King and Macau), India 140, Germany 136, Russia 117 … and Australia 44. Jeff Bezos tops the list (US$177 billion) for the fourth year in row.

If you live in America, obviously, your odds will improve considerably to about 1 in 453,000. In Australia odds are 1 in 591,000. Only 328 women make the list, dwindling their odds to about 1 in 21 million worldwide.

According to the bank Credit Suisse, the total global household wealth – not income, it's calculated as assets minus debts – in 2020 was US$399 trillion. There are 52 million people with a wealth of 1 million US dollars or more. This number includes many people in rich countries who may not consider themselves rich but own their houses outright.

These are world's so-called wealthiest 1%, and they hold 50% of the world's wealth. Forty one percent of them live in the US.

23

An explosion of myopia

The odds of a person developing myopia, also called short-sightedness or near-sightedness depending upon where you live, are about 1 in 5; these odds will shorten, no pun intended, to nearly 1 in 3 by 2020 and 1 in 2 by 2050.

These odds translate to more than 1.45 billion people around the world currently suffering from some form of myopia; the number is estimated to rise to about 2 billion by 2025. Numbers from major East Asian countries, including China, Taiwan, Japan, Singapore, and South Korea are dramatic: up to 90% adults in these countries are myopic. In China alone the numbers are unimaginably high: 223 million people 14 or younger suffer from poor vision.

A myopic person can see reasonably clearly at a short distance, but distant objects appear blurry. Myopia develops commonly in school-aged children, peaking in early adolescence.

What are the reasons for this unprecedented rate of myopia among children and adults? Such rapid changes can't be genetic, says Kathryn Rose of the University of Technology in Sydney, "environmental factors play a huge role in the prevalence of myopia."

Close-up activities like reading and using computers, tablets and smartphones interfere with normal blinking and strain eyes. When used extensively, rather abused, electronic devices can lead to myopia. When your mother warned you about reading in the dark, she was right. Reading in poor light or reading a lot – more than eight hours a day – is not considered good by ophthalmologists and optometrists. They recommend taking frequent 10-minute breaks from near-work and looking at the distance.

In China and many other East Asian countries, probably because of the Tiger-Mom syndrome, children spend most of their day indoors and are hardly ever outdoors. They are so busy with their studies at school, at cram schools and at home that they rarely get a chance to look at the sky. In most countries, school children usually play outdoors during their lunch break; in China, they take a nap indoors.

The best way to prevent myopia in children is to spend more time outdoors. Myopia develops when the eyeball becomes too long and it focuses an image in front of the retina instead of on the retina, which results in a blurred image. Exposure to daylight helps the retina to release a chemical that slows down an abnormal increase in the length of the eyeball.

The World Health Organization warns that untreated myopia is the main cause of vision impairment. When myopia develops early, it has time to develop into something more severe.

Many studies suggest that myopia increases with level of education. "The higher the academic stress, the higher the prevalence and the earlier the onset of myopia," says Maria Liu of the University of California at Berkeley. Is wearing glasses a sign of intelligence? Some studies do show a correlation in myopia and IQ. When it comes to wearing glasses, the odds seem not to be in favor of educated and intelligent people.

24

Baneful bolts of lightning

About 100 lighting strikes take place every second worldwide; that is, over 8 million strikes in a single day. These strikes result in only 100 deaths a year; 20% lightning strikes result in death. Thus, the chances of you not being struck by lightning are reassuringly very, very low (we can say, proverbial 1 in a million). The exact odds of being struck by lightning are not easy to calculate as they are not the same for everyone, everywhere.

In Australia, on average, about 10 deaths and more than 100 injuries are caused by lightning strikes in a year. This means the odds of being struck by lightning in Australia are about 1 in 200,000. In the US, the odds are about 1 in 1.1 million (with odds of 1 in 7.5 million California is at the lower end, odds of 1 in 614,000 puts Florida on the top of the list of US states).

At any time, more than 2,000 thunderstorms occur worldwide, each producing a lightning flash about 5 km (3 mi) long, but only 1 cm (less than half an inch) wide. Each flash carries a current of 30,000 to 50,000 amperes (even your 10-ampere toaster is deadly). Your best chance of avoiding this deadly bolt is to remember the 30-second rule: it takes less than 30 seconds to hear thunder after seeing a lightning flash. When you see a lightning flash, lightning is close enough to pose a threat, immediately rush indoors. If you can hear thunder you, you're within the striking distance of the storm.

Never say that lightning never strikes twice. It does. Some people have been hit by lightning twice and survived. The chances of survival are indeed very, very low (or proverbial 1 in a billion).

The odds of you being struck by lightning may be long, but lightening is the trigger for numerous bushfires in Australia (and wildfires in the US) which not only destroy humans and animals but also built environment and natural ecosystems. A US study points out that lightning strikes may increase by 12% for every degree Celsius of global warming.

Depending upon whether you are a climate alarmist (a term used by Fox News for those who believe in the science of climate change) or a climate skeptic, you may accept or reject this prediction linked to the rise in the planet's temperature caused by the accumulation of greenhouse gases.

25

Bewitching balls of lightning

Lightning balls are nature's rarer phenomena, and they are also least understood. Unfortunately, most of us will never see it. The likelihood of you ever witnessing it is exceptionally unlikely (less than 1% probability).

Ball lightning is always observed during stormy weather. A lightning ball is usually seen as a free-floating, luminous sphere that shines for a few seconds to a few minutes before it either explodes with a sharp bang or flicks out in silence. It can be almost any color, sometimes even a combination, but green and violet are rare.

Its size varies from a small ball to a giant globe several meters in diameter. It may suddenly appear in the air, or even appear holes in the ground, chimneys, sewers, and ditches. It usually moves horizontally in the air (at speeds between 3.5 and 350 km/h or about 2 to 215 mi/h) about a meter (39 in) above the ground but can climb utility poles and then dart along power or telephone lines. It can even dive down chimneys and squeeze through spaces much smaller than its size, but it never changes its size.

It seems cool to the touch, but it may destroy electrical equipment, melt glass, ignite fires, and scorch woods or singe people and animals. Sometimes a hissing or crackling noise can be heard. It may leave behind a sharp and repugnant smell, resembling ozone.

Over the years, scientists have collected thousands of accounts of sightings of ball lightning. In 2002, the Royal Society's journal *Philosophical Transactions* presented a selection of recently reported sightings. One account describes a lightning ball as it entered through an open window in the pantry of a house in Johannesburg: 'It entered the kitchen around the corner then sped out of the kitchen again around another corner and into the passage and the hall where it hit the tin bucket with a clang! Certainly, when we ran to check, the bucket was too hot to pick up and its paint had blistered!'

In another account, a white-grey lightning ball about 80 cm (31 in) in diameter and with the glow of an incandescent lamp of 200 watts bounced on the head of a Russian teacher who was with her friends: "It appeared as if from nowhere. We got frightened, squatted, and connected our heads, creating a circle. The ball suddenly began to move over us in a circle, and it also moved up and down. It was at a height of 50 cm (19 inch) above the ground. Then it "chose" my head and began to jump on it, up and down, like a ball. It made more than 20 jumps. It was as soft as a bubble."

The journal also listed an extraordinarily large – about 100 m (328 ft) in diameter – lightning ball that was caught on color film by a park ranger in Queensland, Australia. It was anchored to the ground and lasted surprisingly long, about five minutes.

If you ever see a lightning ball, take a selfie with it. It would indeed be a rare photograph. In 2012, Chinese scientists took the first ever scientific recordings – video and spectrographs – of ball lightning in nature. The 5 m (16 ft) wide ball rose from the ground following a bolt lightning and travelled horizontally about 15 m (50 ft) before disappearing after two seconds.

26

Ufology

The odds of you spotting a UFO today or any day of your life are zero, yet nearly 1 in 6 American adults report having seen a UFO. And 1 in 3 American adults believes in UFOs. Globally, the odds are 1 in 5 adults.

Aliens live and they live in our midst disguised as aliens. That's what 20% people polled in a global survey believe. The international news agency Reuters 2010 poll of 23,000 adults in 23 countries showed that China and India had the largest percentage of believers (40%) while those least likely to believe in aliens were from Belgium, Sweden, and the Netherlands (8% each). Most of the believers were under the age of 35, and across all income classes. They tend to be men (22% vs 17%).

The term "UFO" (unidentified flying object) was suggested in the mid-1950s by the US Air Force. The term "flying saucer" was not considered accurate, since many sightings had very natural explanations, while others did not.

UFOs are so infrequent that they are unique to most observers. In that sense, they are true encounters of the UFO kind, but they are not encounters with extraterrestrials. "U" in UFO simply means "unidentified;" it doesn't suggest "extraterrestrial."

Most UFO fans believe that extraterrestrial intelligent beings are visiting the Earth. They also believe that governments are covering up this fact because they know it would trigger a panic; governments are afraid of admitting something that is beyond their control.

Most of these UFO fans have also claimed to have seen a UFO. What have they seen? Planes, jets, helicopters, balloons, strange flocks of birds. Unusual light patterns caused by astronomical and meteorological phenomena. Optical illusions caused by smoke and dust. Psychological delusions. Deliberate hoaxes.

Many people who have had vivid memories of close encounters with UFOs are not convinced by these explanations. Scientists say that some UFO events are almost attributable to physical, electrical, and magnetic phenomena in the atmosphere. These events create regions of electrically charged plasma which appear as bright, fast-moving objects to observers. They are probably caused by a meteorite entering the atmosphere, neither burning up completely nor impacting as meteorites, but forming buoyant plasma. Sometimes the field between certain charged buoyant objects forms an area, often triangular, which does not reflect light. This explains why some UFOs are described as black spaceships, often triangular, and up to hundreds of meters in length. The events cannot be detected by radar.

The proximity of plasma related fields can adversely affect a vehicle or person. It has been medically proven that local electromagnetic fields can cause responses in the temporal lobes of the human brain. These result in the observer sustaining (and later describing and retaining) his or her own vivid, but mainly incorrect, description of what is experienced.

27

Lipid years of your mid-life are risky

Globally, if you're over 25 the chances of you having elevated total cholesterol are 2 in 5, or 40%. If you're between the ages of 35 and 55, for every decade you have even mildly elevated cholesterol increases the risk of heart disease by 40%.

"Watch your blood cholesterol level," is the mantra physicians preach to their older patients. Too much cholesterol in your blood can lead to a gradual build-up of fatty material in the walls of your blood vessels. It causes atherosclerosis, or hardening of the artery, which is responsible for a large proportion of heart pains, called angina, or heart attacks.

There are two types of cholesterol: LDL (low-density lipoprotein) is called "bad" cholesterol because it is deposited on the artery walls and clogs them; HDL (high-density lipoprotein) is called "good" cholesterol because it unclogs the arteries. The higher your LDL level, the higher your risk for heart disease. But the higher your HDL level, the better.

When we talk about cholesterol levels, we usually mean total blood cholesterol. In some cases, physicians measure specific levels of LDL and HDL as well triglycerides (another type of fat in the bloodstream). The following figures are for total cholesterol levels in adults; they are approximate and depend on other risk factors such as family history, obesity, high blood pressure, smoking and any history of heart attacks or stroke.

The recommended total cholesterol levels are:

5 mmol/L (193 mg/dl) or less for healthy adults
4 mmol/L (154 mg/dl) or less for those at high risk

28

Volunteering: the odds-on favorite to make you happy

The mathematics of volunteering is skewed (bookmakers don't like it): give a little, get a lot back.

The essence of volunteering distilled by tons of research is simple: by helping others you help yourself

What factors create the protective effects of volunteering? One possibility is that volunteering provides meaning and purpose in people's lives – and a sense of belonging (which is important for older adults as they are prone to social isolation). Such qualities may in turn lead to better health, greater happiness, less stress and depression and a little better sleep.

Studies of older adults have established that people live longer because they volunteer, rather than that people volunteer because they're healthier and hence more likely to live longer. Undoubtedly, better health leads to continued volunteering which leads to improved physical and mental health.

A London School of Economics and Political Science study that examined the relationship between volunteering and measures of happiness in a large group of American adults has discovered that the more time people spent in volunteer efforts, the happier they were. Compared with people who never volunteered the odds of being "very happy" rose 7% among those who volunteer monthly and 12% for people who volunteer every two to four weeks. Among weekly volunteers, 16% felt very happy – a hike in happiness comparable to having an income of $75,000-$100,000 versus $20,000.

To experience health benefits of volunteering, you must be within "volunteering threshold," meaning you must give, on average, 100 hours a year to charity. If you volunteer for more time than this, you will not necessarily gain greater health advantages. Adults 65 and older are most likely to receive health benefits.

But volunteering benefits people of all ages. It's one of the best ways to get your mind off your aches and pains is to get your mind on somebody else. "Helping is a buffer against helplessness, and an affirmation of self-efficacy – I can do this!" says American ethicist Stephen G. Post, the best-selling author of *The Hidden Gifts of Helping*. It does seem that it involves the brain, it involves the immune system, and it probably involves certain hormones, like oxytocin – the compassion hormone, he says. Volunteering also elevates the levels of the body's endorphins and dopamine, the neurochemicals that affect your mood.

Feelings of loneliness are quite common among older people. When loneliness becomes a chronic condition its effect on health can be more serious. These harmful effects include higher levels of anxiety, negative mood, dejection, and stress; and physical effects such as increased risk of high blood pressure, stiffening of arteries, infections, and impaired sleep. An activity which won't clog arteries and will provide ample opportunities for vital face-to-face interactions is volunteering for a cause in which you believe.

Volunteering gives older people a perfect opportunity to leave their "homes full of silence."

29

To turn odds in your favor, stop making your heart work harder

Globally, if you're over 25 the chances of you having raised blood pressure are 2 in 5, or 40%. Men have slightly higher prevalence of raised blood pressure than women. Raised blood pressure is estimated to cause around 12.8% of the total of all deaths worldwide.

High blood pressure, also known as hypertension, is a "silent killer" because people often cannot sense or realize its symptoms. Blood pressure levels have been shown to be positively and continuously related to the risk for stroke and coronary heart disease. In some age groups, according to the World Health Organization, the risk of cardiovascular disease doubles for each increment of 20/10 mmHg of blood pressure, starting as low as 115/75 mmHg.

To pump blood continually to the body via a network of vessels, the heart beats up to 100,000 times a day. Blood pressure is the pressure exerted by blood in arteries. It is measured in millimeters of mercury (mm Hg) and is always expressed in two numbers that represent systolic and diastolic pressures (the systolic number on the top and the diastolic number on the bottom). Systolic is the pressure when the heart muscles contract and blood flows into the

72

arteries. Diastolic is the pressure when the heart muscles relax and the heart fills with blood from the veins.

Normal blood pressure in adults is generally less than 120/80 mmHg. Normal to high blood pressure is between 120/80 and 140/90 mmHg (prehypertension). Blood pressure above 140/90 mmHg is high (hypertension). Blood pressure above 180/110 mmHg is very high (hypertensive crisis; it may require emergency care

30

Get ready to say hello to Martians

The discovery of methane burps shortens the odds of finding life – microbial, not of the little-green-men variety – on the red planet, but the odds still vary from maybe to almost certain.

NASA's Curiosity rover which has been exploring Mars has noticed that the planet is periodically releasing "burps" of methane, suggesting at the possibility of life actively belching it out. The gas is produced by microbial life; on Earth 90% of methane comes from microbial organisms. The Martian bursts of methane could also come from its ice-trapping cages called clathrates in which methane is trapped. But how did the gas get into the clathrates in the first place? Did it come from microbes?

Organic molecules like methane, which contain carbon and hydrogen, are chemical blocks of life. They can also exist without the presence of life.

Scientists have been excited about the prospect of life on Mars – intelligent Martians, not microbes – since the beginning of the 19th century. In the early 1820s German mathematician Carl Friedrich Gauss came up with a brilliant idea to contact Martians. At that time, it was believed that Martians could look at Siberia. To attract their attention, Gauss proposed clearing stretches of forest there to form a gigantic right-angled triangle with squares on each side;

planting wheat in the triangle and leaving squares of trees around it. He believed that his diagrammatic demonstration of Pythagoras' theorem would reveal our presence to our neighbors (how could Martian masterminds miss the mathematical message hidden in Pythagorean "crop circles"?).

A few years later, Johann Joseph von Littrow, an Austrian physicist, worried about the poor visibility of Pythagoras' diagram in the dense Siberian Forest, suggested that the right thing to do would be to dig giant ditches in the Sahara in the form of geometric figures such as triangles, squares and circles, fill them with water, and pour kerosene on top of the water and set them ablaze at night. Even this bright idea was not followed up.

The idea of a blazing Sahara as a beacon to attract Martians' attention appeared again in 1880 when French astronomer Camille Flammarion suggested that, if chains of light were placed on the Sahara on a sufficiently generous scale to illustrate Pythagoras' theorem, intelligent Martians might conclude that there was intelligent life on Earth.

In 1924 David Todd, an American astronomer, convinced the US government to turn off high-powered radio transmitters on 23 August, when Mars came closest to Earth. On that day, when the two worlds were just 56 million km (34 million mi) apart, transmitters were turned off for five minutes before each hour, so that Todd could listen to Martian chatter. Martians were also aware of Earth's closest approach. On that day, they decided to turn off their radio stations and sit in silence.

Intelligent life on Earth is now busy listening to Martian burps, not their chatter.

You don't have to climb it just because it's there

If you do decide to climb Mount Everest and ascend above the base camp, you have around 30% chance of making it. The odds you will not survive the expedition are around 1.5%; coming down from the summit is deadlier than trying to reach it.

More than 4,000 people have successfully climbed Everest since 1953 when Edmund Hillary and Tenzing Norgay became the first people to reach the peak. Better equipment and modern weather forecasting have made the challenge to scale the world's tallest peak (8,848 m, or 29,029 ft, above sea level) much easier than it used to be. But the age is still a critical factor for survival. The chances of success decline rapidly for climbers older than 40. Those older than 60 have only 13% chance of success but have a 25% risk of dying.

Oxygen levels above 8,000 m (26,246 ft) – known as the death zone – are too low for a normal person to survive. Significantly low levels of oxygen can make odd things happen to human physical and psychological states. The deadliest year on Everest was 1996, when 15 people died. Twelve Sherpa guides were killed in 2014. The grim reality of perishing on an Everest expedition is that most of the dead are still there buried deep in snow.

Interpreting global warming forecast

Climate change predictions are not certain events. Climate scientists describe them in terms of levels of certainty or likelihood, which refers to the probability of an event or outcome occurring.

The fifth assessment report of the Intergovernmental Panel on Climate Change (IPCC), released in November 2014 (the sixth assessment report is due in September 2022), represents the latest mainstream scientific opinion on climate change. The report says that it was *extremely likely* that the climate change since the 1950s is the product of human activity. In IPCC-speak, *extremely likely* means that the event has a probability of greater than 95%.

If an event is *virtually certain*, there is a greater than 99% probability that it will occur. In its predictions, the report also uses the terms *very likely* (greater than 90% probability) *and likely* (greater than 66% probability), *more likely than not* (greater than 50% probability) and *unlikely* (less than 33% probability). The terms used in the report for scientists' confidence in the evidence are *very high, high, medium, low,* and *very low.*

Compared to the 2007 fourth assessment, the 2014 report presents stronger evidence of the ways the planet is already experiencing the effects of human-caused changes such as sea-level rise, shrinking glaciers, decreasing snow cover, warmer oceans, and more frequent and intense extreme weather events.

Other trends in the report include:

- It is *very likely* that the number of cold days and nights has decreased, and the number of warm days and nights has increased on the global scale.
- It is *likely* that the frequency of heat waves has increased in large parts of Europe, Asia, and Australia.
- The frequency or intensity of heavy precipitation events has *likely* increased in North America and Europe. In other continents, *confidence* in changes in heavy precipitation events is at most *medium*.
- Global mean sea levels will continue to rise at a rate *very likely* to exceed the rate of the past four decades.
- Increases in intensity and/or duration of draught: *low confidence*.

The report also says that greenhouses gases contributed a global mean surface warming *likely* to be in the range of 0.5 °C to 1.3 °C (0.9 °F to 2.3 °F) over the period 1951 to 2010. Further warming will continue if emission of greenhouse gases continues.

33

Odds are in God's favor ...

**There is a 67% chance that God exists, at least,
according to one scientist.**

To work out the probability, Stephen Unwin, author of *The Probability of God*, uses Bayes' theorem, a 250-year-old formula that is still used to work out the likelihood of events. It is at the heart of everything from genetics to Google search, from health insurance to hedge funds.

The theorem, devised by Reverend Thomas Bayes, an English mathematician, was published posthumously in 1763. It, in essence, says that the probability of a given hypothesis depends both on the current data and prior knowledge. For example, it can be used to work out the chances of failure of nuclear power plant by balancing the various factors that could affect the situation.

In his calculations, Unwin considers six factors: (1) people have sense of goodness; (2) people do evil things; (3) there are natural evils such as earthquakes and cancer; (4) major miracles such as Jesus might have risen from the dead; (5) minor miracles; and (6) people have religious experience.

Renowned evolutionary biologist Richard Dawkins thinks that the oddest case he has seen attempted for the existence of God is the Bayesian argument. "If I were redoing Unwin's Bayesian exercise, neither the problem of evil nor moral considerations would in general would shift me far, one way or the other, from the null hypothesis," he writes in *The God Delusion*.

The null hypothesis is a hypothesis that a person tries to disprove, reject, or nullify. In Unwin's case, the existence and non-existence of God is 50% starting likelihood each.

... but they don't favor Nessie

Still people see what they want to see. The odds that an adult will agree that the Loch Ness Monster will one day be discovered by scientists are around 18%, but the likelihood that people like monster stories is a very high 99.9%. These numbers are as reliable as photographs of the elusive Loch Ness Monster.

Monsters, mythical or real, fascinate everyone. But no creature in modern times has fired the popular imagination as the Loch Ness monster, affectionately known as Nessie. Its alleged home is Loch Ness (the word "loch" is Scottish for lake) in the Scottish Highlands. This 290-m (950-ft) deep lake is one of Britain's largest lakes and, as the prestigious journal *Nature* once scorned, "the underworld of fables.:

The fable of Nessie began in 565 when a giant serpent-like creature jumped out of Loch Ness and lunged at one of the monks accompanying St Columba on his mission to convert Scotland to Christianity. The monster disappeared when the good Irish saint made the sign of the cross. Nevertheless, it never disappeared from the public consciousness, as recurring reports that a monster dwells in the dark waters of Loch Ness continued to tantalize.

It leaped into fame in 1933 when the London *Times* published a detailed story on 51 eyewitness accounts and drawings of the monster collected by Robert T. Gould, a retired navy officer. He described the monster as a 15-m (50-ft) long-necked creature with one or two humps and at least two, possibly four, fins or paddles. As the myth of the monster grew so did the hoaxes and fake photographs. All this hoopla led the famous anthropologist and anatomist Arthur Keith to write in 1934 to the *Daily Mail*: "The existence or nonexistence of the monster is not a problem for zoologists but for psychologists."

If there was indeed a monster in Loch Ness, how did it get there? Believers say that Nessie-like creatures could have been trapped when the lake was cut off from the sea at the end of the last ice age some 20,000 years ago. Sceptics argue that if there has been a monster in the lake for so long there would have to be at least 20 animals in the herd to support continued breeding over centuries. There is hardly any food in the lake to support that many creatures.

All so-called photos and videos of the monster suggest that it might be a plesiosaur. It could be speculated that these marine reptiles have adapted to the cold waters of Loch Ness despite their preference for sub-tropical waters. Plesiosaurs disappeared with the dinosaurs, but it seems believers in Nessie are still living among dinosaurs.

35

Flying on a pilotless plane

The likelihood of people saying 'yes' to flying on a pilotless plane is extremely unlikely, at least for a few decades, especially when they learn that there is around a 6 in 10 tragic chance of three international airline planes crashing in a cluster (within a few days of each other) over a 10-year period.

David Spiegelhalter, the British risk-management expert who appeared in an earlier story, has applied his admirable risk-analysis skills to work out the odds of planes crashing in a cluster. The tragic crashes of Malaysian Airlines flight 17 in Ukraine (on 17 July 2014), TransAsia flight 222 in Taiwan (23 July) and Air Algerie flight 5017 in Mali (24 July) which happened within a space of eight days show that flying can still carry some danger.

David Spiegelhalter's statistical analysis shows that we should not be surprised by such a cluster of tragedies as there is around 6 in 10 chance that we should see such a large cluster over a 10-year period. He says that his analysis may appear cold-hearted but is not to diminish the impact of this tragic loss on the people and families involved.

Driverless trains operate in many parts of the world and driverless cars are expected on the roads soon. But it seems that people are reluctant to board on our pioletless plane. "Safety-wise it seems to make sense – flight crew error has been implicated in about half of all fatal airline accidents," advises New Scientist magazine.

As far as this author is considered, he would fly on a pioletless plane only when self-service checkouts in bookstores are selling only authorless books. Unexpected book in the bagging area. Pioletless plane on tarmac waiting for authors.

36

In America, we trust

But what do Americans believe? Odds are in favor of them believing in God and science, but not necessarily the findings of science. One in 3 believe that the Bible is the literal word of God. The disturbing thing is that 13% of American scientists do not think that humans are responsible for climate change.

Americans believe people behave better if God is watching them
More than 9 in 10 American say "yes" when asked the question do you believe in God. Several studies show that most people associate atheists with distrust.

According to a Gallup Poll, only 54% Americans would vote for an atheist presidential candidate, but 68% would vote for a gay or lesbian candidate. In contrast, more than 90% are willing to vote for a black, a woman, a Catholic or Jewish candidates.

Americans believe in science, not necessarily in its findings
Around 70% of adults in a survey by Pew Research Center say that science has made life easier for most people. However, there is a big opinion difference between what the public believes and what scientists believe.

- *Climate change is mostly due to human activity*. 50% of the public agrees; 87% of scientists agree (94% of scientists say climate change is a very serious problem) – 37-point gap.

- *Evolution.* 65% of the public do not believe in evolution; 98% scientists say they believe humans evolved over time – 33-point gap.
- *Growing world population will be a major problem.* 59% of the public agrees; 82% of the scientists agree – 23-point gap.
- *Building nuclear plants.* 45% of the public agrees; 65% of scientists agree – 20-point gap.
- *Genetically modified food.* 88% of scientists say they are generally safe to eat; 37% of the public agrees – 51-point gap.
- *Foods grown with pesticides.* 68% of scientists say it's safe to eat foods grown with pesticides; 28% of the public agrees – 40-point gap.
- *Childhood vaccines such as MMR should be required.* 86% of scientists say they are generally safe; 68% of the public agrees – 18-point gap.
- *Use of animals in research.* 89% of scientists favour use of animals in research; 47% of the public agrees – 42-point gap.

Proud to be an American

The odds that 25- to 29-year-old American would say that he or she is not proud to be American citizen are 1 in 100. The odds decrease to 1 in 50 for younger adults (18-24 years old). It seems young ones are not so patriotic.

37

Descendants of Genghis Khan

Globally, the odds of a man sharing Genghis Khan's DNA are 1 in 200.

Genghis Khan, the Mongolian warrior, established the largest land empire in history. He died in 1227 but millions of men still bear his legacy in the form strings of DNA.

As mentioned in an earlier story, males carry XY chromosomes, females XX. A child inherits one X chromosome from mother and either the X or Y chromosome from father. An X makes the child female, Y a male. As Y chromosome is only found in men, it can be used to trace a single male paternal line over multiple of generations who fan out over a wide geographical area. Powerful men who father children with multitudes of women tend to leave such successful lineages. Genghis Khan is believed to a have fathered hundreds of children.

In 2003 Chris Tyler-Smith, an evolutionary geneticist in the UK, discovered that 86% of men in 16 populations spanning Asia (and 0.5% of men worldwide) shared identical Y-chromosome sequence. Further DNA evidence showed that their lineage began around 1,000 years ago in Mongolia. More recently, geneticists Mark Jobling of the University of Leicester in the UK and Patricia Balaresque of Paul Sabatier University in France analyzed Y chromosomes of more than 5000 men from 127 populations spanning Asia. Their results, published in 2015, show that Genghis Khan's paternal lineage stood out again.

The evidence of Genghis Khan's influence on global gene pool is still not virtually certain (with odds of 99 in 100) but it's quite likely (odds of 7 in 10).

38

There is a chance of rain tomorrow

What does, say, 40% chance of rain mean?

Weather forecasting is not an exact science and that's why weather forecasts are never certain (probability 100%) or virtually certain (probability 99%).

Rain forecast statements are often expressed as the "chance of precipitation." What weather people call "the probability of precipitation" is the chance of precipitation (rain, sleet, snow, hail and drizzle) occurring at any point you select in the area. Mathematically, probability of precipitation is C x A, where C is the confidence that precipitation will occur somewhere in the forecast area, and A is the per cent of area that will receive measurable precipitation, if it occurs at all. The probability is for a specified period (that is, today, afternoon, tonight, Friday).

When the forecast says that there is 40% chance of rain, the forecaster is expressing a combination of degree of confidence and the area. If the forecaster knows that rain is sure (confidence 100%), they say 40% area will receive measurable rain. If the forecaster's confidence is 50%, the rain, if it does occur, will be measurable rain in 80% of the area.

In Australia, the descriptive terms used for "chance of rain" are: no mention of rainfall (0%, 10%), slight (20%, 30%), or medium

(40%, 50%, 60%), high (70%, 80%) and very high (90%, near 100%). The US National Weather Service prefers: none (0%), slight chance, isolated (10%), slight chance (20%), chance, scattered (30-50%), likely, numerous (60-70%) and categorical, "rain this afternoon" (80-100%).

39

Playing a zero-sum game

In a zero-sum game, each player benefits at the expense of others. The odds of winning or losing are even (1 in 1).

A zero-sum game, in simple words, is a situation where gain or loss of one or more participants equals to loss of gain of other participants. The gains always equal losses, so the sum of the wins and losses is always equal to zero.

Tic-tac-toe, also known as naughts and crosses, and spelt in many ways, is a two-player zero-sum game. As you know, in this pencil-and-paper game for two players, two players take turns to mark either O or X on one of the nine cells in a 3 × 3 grid. The player who first places three identical marks in a horizontal, vertical, or diagonal row wins the game. Though there are 255,168 possible games, you cannot win the game against an expert opponent. An expert player will always draw, as tic-tac-toe is a zero-sum game. Therefore, against an expert player there is 0% chance of winning and 100% chance of drawing.

Some winning tips if you are playing against a novice.

- If you are the first to start, begin with a corner cell.
- If your opponent starts with center opening, counter the move by occupying a corner cell.

- If your opponent occupies any cell (a) next to you, choose the center cell; (b) horizontally or vertically opposite cell, choose the cell that is diagonally opposite to you; (c) any other cell, choose the cell opposite your first cell.

40

Washing your hands, again

About 3 in every 100 people will develop obsessive-compulsive disorder (OCD) at some time in their lives. OCD tends to run in families. If you, or your parents or siblings, have OCD there's around 25 per cent chance that another of your immediate family member will have it.

Mathematicians are known for their eccentricities, but Kurt Gödel's eccentricities are legendary. In his later years, he withdrew from all human contact and received communications only through a door in his office. He was so afraid of food poisoning that he would eat only portions his wife tasted first. After she became too old to do this, one of the greatest logicians of all time refused to eat and died of malnutrition in 1978. Gödel was afflicted with obsessive-compulsive disorder (OCD) which made to him do weird things.

OCD is a brain disorder in which the brain becomes struck on a certain urge or thought. Obsessions also trigger feelings of disgust, fear, or doubt. Some of the common OCE obsessions or compulsions are arranging and ordering objects, washing hands repeatedly and fear of contamination from dirt and germs (towards the end of his life the American tycoon Howard Hughes lay naked in a bed in darkened rooms which he considered germ free).

In his enlightening book, *The Man Who Couldn't Stop: OCD, and the true story of a life lost in thought,* David Adam, a writer and editor at the science journal *Nature,* writes that as an 18-year-old student he spent a night with a pretty girl but didn't sleep with her. The following morning when a friend joked that he might have caught AIDS, the idea lodged in his mind and eventually developed into the incapacitating obsession with AIDS. It took him years to identify it as OCD. *The Man Who Couldn't Stop* is not a self-help book but those afflicted with OCD will find this mix of science, history, and personal memoirs a considerate and compassionate book.

There are no laboratory tests to diagnose OCD. Medication and cognitive behavior therapy can reduce or even eliminate symptoms of OCD.

It's okay to wash your hands again and again during COVID-19 pandemic. A little bit of OCD may keep you safe.

41

It's an odd, odd, odd, odd world

Odds in a world of infinite wonders.

Moon-struck odds

Odds favor man-in-the-moon side to always face the Earth.

We all know about the shape of a human face on the moon's surface. New data from NASA supports the idea that magma from within the moon itself, not an asteroid impact, created the splotches some call the man in the moon.

But why do we always see this side of the moon, not the other side? A study by Oded Aharonson and his colleagues at the University of California dismisses the previous view that man-in-the-moon side permanently facing the Earth is just a coincidence. He says that it may be the result of the Moon being like a "loaded dice." The far side of the moon has a thicker crust, higher mountain, and an elevated topography; and therefore, more gravitation pull. It makes sense that this side should face the Earth.

Four billion years ago the moon rotated much faster and all of the moon was visible from different parts of the Earth at various times. The Earth's pull slowed down its spinning. The rate at which the spinning slowed down determined which side of the moon we wound up with. A faster slow down would have resulted in an even chance of either side eventually facing the Earth. But the moon slowed down more gradually, we wound up being twice as likely to have man-in-the-moon side facing us.

Unlike a coin that has 50:50 chance of showing heads or tails when it's tossed in the air, Aharonson says, the moon was loaded.

Drunk odds
Odds favor drunk trauma victims.

But it's not an excuse to get drunk. A retrospective study of about 8000 trauma victims published in the journal *American Surgeon* reveals that 7% of trauma patients who came in sober died of their injuries, while those who were hurt while drunk only died 1% of the time. A positive blood alcohol level seemed to increase the chances of survival, even after considering other factors such as the age of the patient and the severity of the injury. Avoiding trauma in the first place is still the strategy one might call most sober, advises a report in *Scientific American* magazine.

Musical odds
Romantic music boosts young men's chances with young women

Researchers at the University of Brittany in France exposed 18- to 20-year-old women to either music with romantic lyrics or neutral lyrics while they waited to complete a taste test with a 20-year-old male research assistant posing as another student volunteer. When during a break the research assistant asked the women participants for their telephone number, 52% percent of the women who heard romantic music said yes compared with 28% who heard neutral music.

Missing odds

The odds of your lost pet cat returning home are 3 in 4; for lost dogs, the odds are nearly 95 in 100.

This good news for pet lovers comes from the American Society for the Prevention of Cruelty to Animals (ASPCA). Can we extrapolate from these odds that they apply worldwide? The answer is "yes" if the cats and dogs in the ASPCA survey were only Stars and Stripes waving patriots. We consider cats and dogs as global citizens, and all over the world they usually come home all by themselves.

In Kafka on the Shore, a novel by the world's best-known living Japanese novelist Haruki Murakami, Nakata is a finder of strayed pet cats. He can even talk to cats. That's magic realism. The following is reality.

In an experiment researchers assigned participants in one of the three conditions: a pet nearby, simply thought about a pet and no pet presence. Those who had their pet in their room or in their mind identified more goals and showed confidence in achieving them. In a second experiment the participants performed a distressing mental task while their blood pressure was measured. Participants with their pet in the room or in their mind had lower blood pressure than the participants with no pet presence.

If you do lose your cat or dog, just think about them, and not worry about the odds of finding them.

Dishonest odds

If you want to catch someone cheating or lying, you'll raise your chances in the afternoon (only if they are usually honest and ethical to begin with)

As the day wears on our ability to show self-control to avoid cheating or lying wears down. We are more honest and ethical in the morning than in the afternoon, say American researchers Maryam Kouchaki and Isaac Smith.

They studied the phenomenon, which they have dubbed "morning morality effect," by giving a series of tests to 327 college-age men and women. In one of the tests, students were shown various patterns of dots on a computer and were asked to identify whether the dots were displayed on the left or right side of the screen. They were paid small amounts of money based on which side they determined had more dots, but the amount was ten times more for selecting the right over the left.

They reported their own scores, giving them an opportunity to cheat. People who participated in the morning session were less likely to cheat than those who took part in the afternoon session.

Another test was designed to test students' moral awareness in the morning as well as in the afternoon. They were asked to fill spaces in word fragments such as "– –RAL' and 'E– – –C– –." In the morning, students were more likely to fill in the words 'moral' and 'ethical' while in the afternoon they tended to choose 'coral' and "effects." Something as mundane as the time of day can lead to a systematic failure of good people to act morally. This effect is linked to our body's energy levels, which inevitably declines as the day goes on.

"Unbanked" odds
Globally, the odds of an adult having an individual or joint bank account are 1 in 2

As you would expect in the developed world 98% adults have

accounts, and in the developing world the figure is less than half, 41%. More than 2.5 billion people around the world don't have a bank account. These are the world's "unbanked" poor.

Twin odds

What are the odds of twins born in different years?

What does it really mean? One of the twins is born before midnight on 31 December and the second one after midnight on 1 January. The odds of a live birth of twins born in different years are 1 in 59,000, according to freakonomics.com. The calculation apparently applies only to the US.

Genetic odds

When it comes to numbers, the odds are in your genes.

If you have easily understood the number, you have read so far in this book, you have an inherent sense of numbers. Studies point out that some people are born with a naturally better sense of numbers than others. An earlier study of an Amazon tribe that received no mathematics education showed that their scores on a standardized number system test were like those in an educated French population.

Emoji odds, American style

Emoji-using American singles have better chances of having sex than those who abstain from emoji use

This revealing information comes from the dating website, match.com. Whether you are a man or woman, the odds are even that you would say emojis make it easier to express feelings. Most used emojis are:

Wink 53%   Smiley 38%   Kiss 27%

This old deal is still worth making

This deceptively simple probability problem is so counterintuitive that it leaves most people scratching their heads.

Let's Make a Deal is a television show running on and off and in various formats since 1963. The famous Monty Hall problem is loosely based on the original show and is named after its host, Monty Hall. If you go on the show you are asked to choose one of the three doors. Behind one door is a shiny new car; the other two have goats. Your odds of winning the car are 1 in 3. Let's say you choose door A.

After you have chosen the door, Monty Hall who knows what's behind the doors, opens door B to reveal a goat. He now asks you whether you would like to stay with your original choice or switch to the other unopened door. Should you stick with door A or switch to door C?

Most people think that with two unopened doors remaining your chances of winning are the same, 50-50, so switching makes no difference. Surprisingly, switching door doubles your chances of winning the car. Your odds of winning the car are now 2 in 3.

The Bayes theorem, mentioned in an earlier story, says that the probability of a given hypothesis depends both on the current data and prior knowledge. When you were asked to pick a door, you knew that the odds were 1 in 3 for any given door. When Monty Hall opens door B, you update your prior knowledge with new data that door B has a goat. The odds you guessed right that the car is behind door A remain 1 in 3. If you guessed wrong the odds are 2 in 3. The odds are also 2 in 3 if you guessed it wrong that the car is behind door C. So, to improve your chances of winning from 1/3 to 2/3 you should switch to door C. The following table lists all choices.

Door 1	Door 2	Door 3	Stick	Switch
CAR	goat	goat	CAR	goat
goat	CAR	goat	goat	CAR
goat	goat	CAR	goat	CAR
			Winning odds 1 in 3	Winning odds 2 in 3

No probability problem is as good as the Monty Hall problem in fooling people. In 1991 Marilyn vos Savant, an American newspaper columnist listed in the Guinness Book of Records Hall of Fame for "Highest IQ," published the problem's uncanny solution that the contest should switch doors. But her 228-point IQ and impressive analysis of the problem failed to convince her readers: more than 10,000 of them sent irate letters, the great majority disagreeing with her. *The New York Times* reported at the time that the most vehement criticism came from mathematicians and scientists, who had alternated between gloating at her ("You are the goat!") and lamenting the nation's innumeracy.

The problem even fooled Paul Erdös, one of the greatest mathematicians of the 20th century. He remained unconvinced until he was shown a computer simulation confirming the predicted result. You would not expect a simple puzzle leaving a mathematician who had published more than 1500 papers (a great mathematician may publish 50 papers in a lifetime) scratching his head.

But the problem never fooled a computer. A computer programmed to play Monty Hall game played it 100,000 times. A contestant who always switched doors won the prize 66,841 times, while a contestant who never switched won 33,329 times. Marilyn vos Savant was right.

A sock performance: From three doors in a bright studio to your socks drawer in a dark room. In the drawer, there are two black socks and one blue sock. If you take out two socks, one after another, what are the odds of getting two matching black socks? Think for a while before you look up the answer below.

Answer: The chance of getting a black and a blue sock (mismatching) is double that getting two black socks (matching). Now you know our intuition about the odds of things happening is never reliable

43

Not stopping at the red light

Color blindness (or color vision deficiency) affects about 1 in 12 men and 1 in 200 women in the world. Color blind people are not blind to all colors; most have red-green color vision deficiency. Only 5% of people who are color blind have blue-yellow color blindness. Complete color blindness, when people are unable to see any color at all, is extremely rare, affecting about 1 in 100,000 people worldwide.

No, red-green color blind people do not fail to stop at the red traffic light. They can see things as clearly as others, but they are unable to fully see red and green colors. They can easily identify the difference between the red and green traffic lights. Those who are totally color blind check the position of the light that is lit (red is always at the top).

Red-green color blind people do not mix up red and green colors; they mix up all colors that have some red or green as the part of the whole color. The oft-repeated story of the English chemist John Dalton's color blindness provides a classic example of such a mix up. The story goes that Dalton gave his mother a pair of stockings as a birthday present. "You have bought me a very fine pair of stockings, but what made you fancy such a bright color?" she asked. "I can never show myself at meetings in them.

They're as red as cherry, John." Much distressed by his mother's remark, he asked his elder brother who, like Dalton, thought the stockings were a drab dark-bluish color. When their neighbors also remarked "very fine stuff, but uncommon scarlety," Dalton realized that he and his brother suffered from some genetic defect and decided to investigate color blindness. He published the first-ever scientific study of color blindness in 1798.

Color blindness is indeed a genetic defect. It's caused by deficiency in cone cells in the retina which detect colors, a fact Dalton didn't discover (he thought the insides of his eyeballs were blue). The genes that lead to red-green color blindness are on the X chromosome; this fact makes color blindness rare in women. Males carry XY chromosomes, females XX. A functional gene on only one of the two X chromosomes is sufficient to produce normal color vision.

There is no medical treatment to cure color blindness. However, certain types of colored filters or contact lenses can offer some help to make up for color vision problem.

44

Probability is, you probably don't exist

Only if you believe in an American author's answer to the question: What are the odds that you came about and exist as you today?

Before we present calculations of Ali Benazir, author of *The Tao of Dating* and many other books, we advise you to take them with a grain of salt, rather quite a few grains of salt.

His journey to the answer begins with the odds of your father meeting your mother: 1 in 20,000. When he multiplies this figure with the odds of them staying together (1 in 2,000) the answer comes to 1 in 40 million. Benazir then presents the odds of right sperm meeting the right egg to make you (and not your sibling) as in 1 in 400 quadrillion (that is, 4 followed by 17 zeroes).

We are just getting started, he warns, don't get carried away.

Because your existence here now and on the planet presupposes another supremely unlikely and utterly undeniable chain of events: every one of your ancestors lived to reproductive age, going back to not just the first *Homo sapiens* but all other ancestors in the chain of evolution all the way back to the first single-celled organism 4 billion years ago. If we stick to human generations only, the total number of generations would be about 150,000. The odds of every one of your human ancestors reproducing successfully are 1 in $10^{45,000}$ (that is, 10 followed by 45,000 zeroes).

The odds of the right sperm, meeting the right egg 150,000 times are 1 in $10^{2,640,000}$ (that is, 10 followed by 2,640,000 zeroes). He arrives at this number by multiplying 400 quadrillion 150,000 times.

To get the final odds he multiplies $10^{2,640,000}$ x 40 million x $10^{45,000}$. The odds are 1 in $10^{2,685,000}$.

If you compare it with the puny number of atoms in the universe, 10^{80}, you will agree the answer is basically zero.

His advice is: Now go forth and feel and act like the miracle that you are.

Thank you, Ali Benazir. But we know the probability of existing something that exists can only be described as absolutely certain 100%. In a world ruled by chance and chaos we do exist. Absolutely, not probably.

45

Playing the cosmic lottery

**Our guide to laying odds on the apocalypse. But please
don't bet your bottom dollar on these odds (even if you
do, odds are such that you won't be there to honor your
bet).**

Our planet has survived for 4,600 million years; no celestial
cataclysm has succeeded in disintegrating it into cosmic dust.
Similarly, life on this puny planet has dodged total extinction many
times for 3,500 million years when battered by killer comets and
asteroids from the sky and deadly megaquakes and supervolcanoes
from the deep. For nearly 2 million years, humans have survived
ice ages and other natural catastrophes. It's true that civilizations
exist by geological consent and when nature suddenly withdraws
this consent, rumbling earthquakes or oozing volcanoes wipe out
vast swathes of human societies, not humankind.

The word apocalypse – it comes from a Greek word meaning
uncover or reveal – now means destruction on a catastrophic scale,
and Armageddon is synonymous with the end of the world.
Whatever we do, it's inevitable that the history of the world will
come to close some day. That day could be billions of years away
or it could be tomorrow.

Our fascination with prophecies about the end of the world (or
at least, the world without us) is endless. Here're some prophecies
of the scientific kind (two apocalyptic events – asteroid impact and

nuclear war – have been discussed earlier).

Solar superstorm
Odds of happening in the next 15 years: 1 in 20.

Solar flares are huge bursts of energy which come from sunspots. They produce waves of ionized matter that travel with the speed of millions of kilometers per hour through space. When solar storms hit the Earth's atmosphere in the layer where communications signals travel, they can have many disastrous consequences. They can knock out satellites, power grids and communications system. In our highly electronically connected world, a solar superstorm can cause a devastation of cataclysmic proportions.

The number of visible sunspots varies in a regular cycle, known as the sunspot cycle, reaching a maximum about every 11 years. During a solar maximum – the last one occurred in 2014 – there is marked increase in the number of sunspots and solar flares.

Out-of-control global warming
Odds of happening in the next 200 years: 1 in 2.

Widespread melting of snow and ice is raising global sea level which would erase coastal cities and make refugees of hundreds of millions of people. Uncontrolled global warming could set in motion the irreversible melting of ice sheets in Greenland and West Antarctica. Scientists agree that these sheets are now melting at an unprecedented rate.

Supervolcano
Odds of happening in the next 1,000 years: 1 in 100.

In the past 6 million years, there have been on average two

supervolcano eruptions per million years. The most recent super-eruption occurred 26,500 years ago.

Volcanoes are fickle. They can erupt without any warning spewing enormous amounts of ash and lava which could significantly affect global weather. It is difficult to imagine how a volcanic eruption in one part of the world can disrupt the world's climate. But there is a recent example: the Tambora volcano in Indonesia in 1815 injected so much dust in the upper atmosphere that it led to a worldwide cooling the following year. There were cold spells, frosts, and crop failures in parts of Europe and North America and 1816 is still remembered as "the year without summer." It was the largest volcanic eruption known in the past 10,000 years.

A supervolcanic eruption would be at least 20 times massive than the Tambora eruption.

A nearby gamma-ray bust
Odds of happening in the next 100 million years: 1 in 15.

When a massive star is sucked into a black hole it produces bursts of energy that are detected millions of light years away as gamma rays. Some of these bursts also reach our skies but they are very faint and last only for a few moments. Gamma-ray bursts have detected so far are all immensely far away – half-way across the Universe. A nearby blast of gamma rays would trigger chemical reaction in the Earth's atmosphere, which would wipe out the entire ozone layer. The catastrophe would kill all but the most well-protected or radiation-resistant species. Some scientists believe that a gamma-ray burst caused a mass extinction 440 million years ago.

A passing star ejecting the Earth from its orbit
Odds of happening ever: 1 in 100,000

Computer modelling by US scientists Fred Adams and Greg Laughlin of interactions between stars passing by our Solar System shows that since the Earth is so close to the Sun being ejected by direct encounter with a star is highly unlikely. Their odds of such a dramatic apocalypse are 1 in 2.2 million. Other planets are highly vulnerable. They say that the odds are 1 in 100,000 that a star will alter Jupiter's orbit, making it pass so close to the Earth that it could fling the planet into deep space or crush into the Sun.

46

The likelihood of global risks

The global risks described below might not be as apocalyptic as the events described in the previous story, but they have potential to cause untold harms in our world. Almost 900 experts who have evaluated these risks did not give the odds of them happening but described them on a probability scale of 1 (very unlikely) to 7 (very likely).

A global risk, according to the World Economic Forum, is an uncertain event or condition that, if occurs, can cause significant impact for several countries of industries within the next 10 years. In 2015, the World Economic Forum listed the following 10 global risks in terms of likelihood.

1. Interstate conflict with regional consequences (geopolitical; likelihood roughly 5.75)
2. Extreme weather events (environmental; 5.5)
3. Failure of national governance (geopolitical; 5.5)
4. State collapse or crisis (geopolitical; 5.25)
5. High structural unemployment or underemployment (economic; 5.25)
6. Natural catastrophe (environmental; 5.25)
7. Failure of climate-change adaptation (environmental; 5.25)

8. Water crises (environmental; 5.25)
9. Data fraud or theft (economic; 5.25)
10. Large-scale cyber-attacks (economic; 5.25)

Pick anyone out of five – and you're a winner

Globally, if you're over 25 the chances of you having raised blood glucose level are about 1 in 12. One in two people with diabetes does not know they have it. Every seven seconds one person dies from diabetes.

The cause of type 1 diabetes, which is caused by deficient insulin production, is not known. It's not preventable and requires daily administration of insulin.

Type 2 diabetes results from the body's ineffective use of insulin. 90% of people with diabetes around the world have type 2 diabetes which is largely the result of excess body weight and physical inactivity.

Researchers have discovered that dealing with any of the following five lifestyle risk factors can lower the risk of developing diabetes by about a third: obesity, a poor diet, lack of physical activity, cigarettes and too much alcohol.

A report published in the *Annals of Internal Medicine* says that researchers followed the health of more than 200,000 adults aged 50 to 71

over 11 years. The risk of developing diabetes over the study period was about 10% for men and 8% for women. But each healthful change in behavior lowered the risk of developing diabetes by as much as 39% in women and 31% in men.

Children: less sleep means more fat

Children who sleep less increase their odds of becoming overweight.

Not getting enough sleep disrupts hormone levels which may lead to excessive weight gain. This is the result of a meta-analysis of 11 studies that looked at children's sleep duration and their body mass by researchers at John Hopkins University in the US. The recommended sleep duration for children older than 10 is 9 hours. Children between 5 and 10 need 10 hours, and younger than five need 11 hours.

The increase in odds a child being overweight if they sleep less are:

- 1 hour: 43%
- 1 to 2 hours: 60%
- More than 2 hours: 92%

Teenagers: hopeless at perceiving their weight accurately

Almost 1 in 3 male teenagers inaccurately perceive their weight.

This result comes from a study of 597 teenage boys aged 13 to 16 years. They predominantly came from middle-class families of Spanish origin from Barcelona and surrounding regions, but the group also included boys of Latin American, North African, European, and Sub-Saharan origins. One study alone cannot turn odds of 1 in 3 into a worldwide figure, but it does provide an interesting insight into the thinking of teenage boys. Read on.

Universitat Autònoma de Barcelona researchers asked the teenagers what they thought their weight level was – underweight, normal weight, slightly overweight, very overweight – and if they did anything to control weight in the past year. The researchers found that up to 28% of the boys inaccurately estimated their weight. Among those with low weight, 43% overestimated their weight, while 86% of those who were obese underestimated their real weight. Among those who were slightly overweight, 40% believed they had a normal weight. Most boys with normal weight (85%) were accurate in their perception of their weight.

The researchers believe underweight boys who perceive themselves as having a normal weight might have incorporated an ideal of male beauty based on a thin but toned body. In the case of overweight boys who consider their weight normal, it may be because they perceive messages on obesity as something socially undesirable. For this reason, they do not accept the fact that they are overweight.

50

Adults: it's fat, not fiction

**Worldwide obesity has more than doubled since 1980.
Or, we can say the odds of becoming obese have halved.
Around 13% adults aged 18 years and over are obese
worldwide (in Australia, the UK, and the US around 1 in
4 people are obese; in China and Japan around 1 in 20.)**

Knowing your body-mass index (BMI) can give you an idea of how healthy your weight is. BMI is a measurement of body mass based on a person's height. It is calculated by dividing weight in kilograms by the square of height in meters (kg/m^2). Google "BMI calculator" to calculate your BMI online; you only need to key in your weight and height.

The World Health Organization defines a BMI greater than or equal to 30 as obesity. A BMI greater than or equal to 25 is considered overweight. The healthy BMI range for adults is 18.5 to 24.9 (less than 18.5 is underweight and possibly malnourished). Most of the world population live in countries were overweight and obesity kills more people than underweight.

The basic cause of obesity and overweight is an energy imbalance between calories consumed and calories expended. The global trend of eating too much energy-dense junk food and doing too little physical activity is driving this epidemic.

51

Toddlers understand probability, do you?

Children as young as age two intuitively use mathematical concepts such as probability to help make sense of the world around them.

Let's start with you. If you have never played Go Fish card game, here's how you play it. The dealer deals five cards from a standard 52-card deck to players. The remaining cards are spread out face down in a disorderly pile called pool. The goal is to collect sets of four cards of the same rank. When your turn comes to play, you ask your opponent for a matching card. If the opponent doesn't have the card, they say "go fish." You then pick up a card from the pool. Whenever a player has four cards of the same rank, those cards are discarded. The game ends when one player's hand is empty, or the pool is depleted.

To play this game with winning strategy you need to understand probability. For example, if you have three queens and one jack, would you ask for a queen or a jack? If you answer is queen, you don't understand probability. You have only one chance to get a queen (there is only one queen left) but three chances to get a jack (three jacks are still out there).

In a deck of 52 cards, you have 1 in 4 chances of getting a spade (or heart, diamond, club). The calculation is simple: 13 out of 52, or 1 in 4. The odds of randomly drawing a king of spade are 1 in 52. These odds keep on changing as you draw cards from the deck. If the first card you picked up wasn't a spade, there are still 13 spades in the deck which has now only 51 cards. Your odds of drawing a spade have now improved to 1 in 51. With each draw the odds would continue to improve (or shorten, as the bookmakers say).

Children learn many skills by simply watching people around them. In a University of Washington study led by Anna Waismeyer children watched as a researcher played a cause-and-effect game. Children were delighted when the researcher placed a wooden block on to a small box to activate a nearby marble-dispensing machine. One block activated the machine two-thirds of the time, and a differently colored and shaped block triggered the machine only one-third of the time. The children watched the researcher playing the game 12 times using different blocks.

When children played the game themselves, 23 out of 32, or 72%, of the children eagerly put the block with the greater success rate. To make sure that children were making choice based on probability – the better two out of three rate – or frequency, the researchers ran the game again on a separate group of toddlers.

This time probability varied, with one block activating the marble four out of six times (two-thirds probability) and the other block activating the marble machine four out of 12 times (one-third probability). This time 22 out of 32 children picked the more successful block, showing that they were using the difference in probability to their advantage.

The current way of teaching probabilities to children relies on fractions and decimals. Waismeyer hopes that teaching probabilities would be easier if we had our teaching mesh with or build on the intuitive ways children think.

52

They say the house always wins

Someone must win, but the odds are stacked against gamblers.

Casino games such as slot machines, craps, roulette, and baccarat are games of pure chance. In these games, you bet on events that are both random and independent and you have no real chance to make a profit in the long run. In commercial gambling the house always has the edge; it achieves it by paying back less than the true odds of winning.

The law of large numbers says that if you repeat a random event such as tossing a coin or rolling a dice a very large number of times, the relative occurrence of the event will gradually come to reflect the true probability of the event. The law of averages – things average out over time – supports the same phenomenon.

For example, more times you toss a coin, closer you get to the true probability of equal number of heads and tails. To skew the law of large numbers in its favor a casino makes sure that each game is weighted slightly in its favor. To achieve this edge, it relies on the probability theory to calculate the expected payout for each game. It's not easy for a player to determine the house edge from merely playing the game.

In roulette, a popular casino game, a ball spins around the rim of a wheel in the opposite direction to the direction of wheel spinning. The American roulette wheel consists of 38 numbered slots: numbers 1 to 36 (alternatively red and black), 0 and 00 (which are green). When the wheel is spun, the marker ball is equally likely to land on any one of the 38 spots. If you bet one chip on number 15 and the ball comes to rest on 15, you will get back 36 chips. The probability of winning is 1/38, but the payout is 36/38. The house edge on the bet is 1/19 or 5.26%. The green slots represent the house's profit per bet (2 chips for every 38 chips bet.)

The European roulette wheel doesn't have 00 green slot, making the chances of winning a little bit better. It gives the house 2.7% edge.

Whether it's American or European, the odds are stacked against you. The house makes money because, on average, you lose money. Even Russian roulette, the deadly game of chance, offers better odds: if a revolver holds six rounds, the odds are 1 in 6. Not suggesting that you play with a gun, even for a nanosecond.

The Wizard of Odd website (*www.wizardofodds.com*) has comprehensive information on casino games, including the odds and house edge for various games. The website is run by Michael Shackleford, a professional actuary and Adjunct Professor of Casino Math at the University of Nevada, Las Vegas.

Spending time at this website is unlikely to dispel the erroneous idea that outcomes for future bets are predictable from those of previous one. The idea leads gamblers to believe if they have been losing, they are more likely to win in future. This is known as gamblers' fallacy: the belief that the chances of something happening with a fixed probability become higher or lower as the process is repeated. If a coin is tossed repeatedly and heads comes up larger number of times than expected, a gambler might predict that the next coin is more likely to land with tails up.

Juemin Xu at University College, London has analyzed 565,915 sports bets made by 776 online gamblers. The analysis reveals that hot hand does really exist. The hot hand – if you have been winning, you are more likely to win again – is caused by gamblers believing in gamblers' fallacy. After a lucky win, the best strategy is to leave and take a cold shower to cool down that damn "hot hand." Chanting the mantra "the odds of winning are the same on every throw of dice" would also help in curing gamblers' fallacy.

53

The price is just right

When something costs $100, a rounded number, the odds are that your decision to buy it will be driven by your feelings, but if the cost is $98.76, a non-rounded price, you will use reason to work out whether it is a good buy.

If you are in a bookshop and pick up a book priced $20, your mind would be working in a different way thinking in terms of round numbers such as $18, $19 or $21 than if it is priced $19.95, $19.98 or $19.99.

A price just below the next dollar increment gives you a precise mental anchor, a starting point for psychological maneuvering to follow. Several studies support the intuitive idea that we simply ignore the cents and round them down. The left-digit effect, as psychologists call it, makes the left-most digit more influential on our perception of price than the digits on the right.

Two Singapore researchers, Monica Wadhwa and Kuangjie Zhang, have now found why shoppers deal with pricing information differently when prices feature rounded numbers as opposed to non-rounded numbers: a rounded price encourages consumers to reply on emotions when evaluating products, while a non-rounded price encouragers shoppers to rely on reason. "When a purchase is driven by feelings, rounded prices lead to a subjective experience of feeling right," they write in the *Journal of Consumer Research*.

Their study shows that the mere roundness of a price number could significantly influence consumer preferences, depending on whether a purchase is driven by feelings (for example, a champagne priced at $50 rather than $47.98) or has a more utilitarian purpose (for example, buying a calculator priced at $8.99).

Well, this study (the first of its kind, as the researchers claim) was obviously done under "laboratory" conditions. You can test its result in the "real world" when you are out shopping for a diamond ring (priced in round numbers such as $500 or $600) for your partner or shopping for a utilitarian toaster for your kitchen (priced in non-rounded numbers such as $19.99 or $18.76).

While we are on consumer predictions, here's another interesting idea: Do categories matter when predicting the lottery or stock market? People are more likely to make prediction about something when it is grouped in a large category, say US researchers Mathew Isaac and Aaron Brough in the *Journal of Consumer Research*.

In one of their studies, they gave the participants lottery tickets either printed in blue or yellow colored-paper. Most of the distributed tickets were blue. They asked participants to write their name on the back of the lottery ticket and indicate whether they would wager additional money on winning the overall lottery. Despite equal odds of any ticket being drawn, participants holding blue tickets were willing to wager an average of 25% more money than the participants holding yellow tickets.

Category size do impact our perception of risk and probability.

This is age-old wisdom and is encapsulated in the old idiom: sometimes we can't see the forest for the trees (Anglophiles are welcome to replace "forest" with "wood").

54

Playing poker with a computer

The odds are that this computer program will beat anyone. The chances of drawing a royal flush in five-card poker – ace, king, queen, jack and ten of the same suit – are 4 in 2,598,960 or 1 in 649,739 (a royal flush can be formed four ways, one for each suit).

In five-card poker, the most common form of poker, the total number of possible hands a player can have and various tactics such as folding (throwing away the hand), calling (matching the opponent) and raising the bet pushes the total number of different situations to 13.8 trillion. To get there, each of us on the planet would have to play nearly 4,000 hands of poker. A task made in heaven for computers.

In poker, there is absolutely no perfect hand or strategy, says Michael Bowling of the University of Alberta in Canada. Unlike chess or checker, a poker player doesn't always know the past moves of other players. Also, a player can win hand when the other players fold.

Keeping all these probabilities in mind, Bowling and his team has devised a computer program that simulates billions of hands of poker. To run the program, they used 4,000 central processing units for two months, equal to about 1,000 years of computing time. The results alone took up some 15 terabytes of computer storage. Still the program has its limitations: it only works when the computer is playing against a single player.

The University of Alberta researchers didn't devise their program to simply play poker. Though their program plays poker with a poker face, it could have diverse uses, ranging from national security, to tracking fare evasion on transit systems, to making decisions about medical treatments.

If you play poker and love reams of data, you may find some meaning and strategy – and hidden luck – in various hand combinations and odds listed below. In this table, the number of distinct hands adds up to 7,642; and the total number of hands 2,598,960.

Hand	Number of distinct hands	Number of ways to draw the hand	Odds
Straight flush (five cards in numerical order, all of the same suit), but not royal	9	36	1 in 72,192
Four of a kind (four cards of the same rank and one side card)	156	624	1 in 4,164
Full house (three cards of the same rank and two cards of a different matching rank)	156	3,744	1 in 693
Flush (five cards of the same suit but not in	1,277	5,108	1 in 508

sequence), but not straight			
Straight (five cards in sequence)	10	10,200	1 in 254
Three of a kind (three cards of the same rank and two unrelated side cards)	858	54,912	1 in 46.3
Two pair (two cards of the same rank, two cards of a different matching rank and one side card)	858	123,552	1 in 20
One pair (two cards of a matching rank and three unrelated side cards)	2,860	1,098,240	1 in 1.36
High card (any hand that does not meet any of the conditions listed above)	1,277	1,302,540	1 in 0.99

The law that kicks the bucket

Your chances of dying during a given year doubles every eight years (until the chance becomes a reality).

British mathematician Benjamin Gompertz worked as an actuary; a profession involved in calculating the financial impact of risk. In the 1820s he became interested in how he can model the probability of people living to a certain age, if nothing unexpected happened to them. After comparing mortality data in different age groups across four cities in England, he concluded that mortality increases exponentially as age increases. In other words, a person becomes more likely to die while growing older.

The Gompertz law of mortality, encapsulated in an equation, is a highly successful law to model mortality of humans. Real-world data collected by the US Center for Disease Control follow this law almost perfectly. The "Gravity and Levity" blogger explains the law as it applies to him:

> For me, a 25-year-old American, the probability of dying during the next year is a fairly minuscule 0.03% – about 1 in 3,000. When I'm 33 it will be about 1 in 1,500, when I'm 42 it will be about 1 in 750, and so on. By the time I reach age 100 (and I DO plan on it) the probability of living to 101 will only be about 50%. This is seriously fast growth – my mortality rate is increasing exponentially with age.

56

Our fascination with shark attacks

On average, every year around 75 shark attacks on humans are reported worldwide, of which about one-tenth are fatal. Globally, the odds of a person dying in a shark attack in a year are minuscule 1 in 1 billion, but the chances of dying from a falling coconut are much higher: 1 in 47 million. Wait, these figures are dependent on geography.

If you live high up in the Himalayas, your odds of dying from a shark attack or a falling coconut are almost zero. But if you happen to live in Florida and love swimming in New Smyrna Beach, known as the shark capital of the world, you are likely to be within a few feet of a shark. On average, your odds of being bitten by a shark are 1 in 2,000. Overall, the odds a person will die from shark attack in a year in the US are around 1 in 251 million. And more than 90% attacks worldwide are on males.

In Australia, in the past 100 years there were 808 shark attacks, of which 173 were fatal. This figure translates into the odds of roughly 1 in 139 million for an Australian dying from a shark attack in a year. Compared with the US, it's riskier living in Australia, shark-wise. Not otherwise, rest assured.

Most shark attacks occur less than 30 meters (100 ft) from the shore mainly around popular beaches in North America (especially

Florida and Hawaii), Australia and South Africa, according to National Geographic Channel. Falling coconuts are dangerous only when you are under a palm tree, according to common sense.

Another nugget of information from National Geographic Channel: For every human killed by shark, humans kill approximately 2 million sharks. Do sharks know these odds?

Arguably, our fascination with shark attacks began with the thriller movie Jaws (which begat Jaws 2); but most shark victim are not dragged around in circles as you saw in the movie. There are three types of shark attacks, according to Florida Museum of Natural History: hit-and-run-attacks (the shark will "attack" out of curiosity mistaking humans for seal), sneak attacks (which happen without warning and are the result of aggression) and bump-and-bite attacks (the shark will repeatedly circle and then bump into the victim aggressively). The best strategy to avoid a shark attack is to swim in a group as sharks usually attack lone individuals and not people in a group.

The worst shark attack in history happened on 30 July 1945. The American warship, USS Indianapolis was on a secret mission to deliver parts of the first atomic bomb to a naval base in the Pacific Island of Tinian, when a Japanese torpedo hit it just after midnight. The ship sank in just 12 minutes. Of the 1,196 men abroad, 900 made it into the water alive. Survivors bobbed in the water, in their life jacket or life rafts which were scarce. The first night the sharks focused on the dead bodies in the sea. They then turned their attention to the living.

Loel Dean Cox, who was a 19-year-old seaman on USS Indianapolis, told BBC News in 2013; "In that clear water you could see the sharks circling. Then every now then, like lightning, one would come straight up and take a sailor and take him straight down. One came up and took a sailor next to me."

Up to 150 sailors died from shark attack before the USS Doyle arrived on the scene and helped to pull the last survivor from the water.

57

What are the odds of finding a four-leaf clover?

It's a widely Googled question on the internet. Why? As legend has it, finding a four-leaf clover supposed to be a sign of good luck. Yes, we have the odds: around 1 in 10,000. And also, for finding an even rarer five-leaf clover: around 1 in 1,000,000.

Each leaf of a four-leaf clover has a specific meaning: faith, hope, love and luck. No one seems the meaning of the rarer fifth leaf.

Instead of combing through a field of clover in hopeless pursuit of a four-leafed one, *Scientific American* magazine suggests a better strategy. Around 200 clovers can be found in a 60-square cm (24-square in) a plot of clover-growing grass or field, which means, on average, a space of about 1.2-square m (4-square ft) should contain a four-leaf clover.

An Australian woman didn't have to go through all this scientific mumbo jumbo to find her good-luck charm. In 2014 she just plucked 21 four-leaf clovers from her garden in Glen Alpine, near Sydney. Now, for four-leaf clover fans the question is: What are the odds of such an event happening? Sorry, we don't' have the answer. But we can tell you that a person who finds a four-leaf clover will meet a future lover on the same day.

Don't stretch this idea too far: finding 21 four-leaf clovers doesn't mean the lucky woman also found 21 future lovers on the same day. Anyway, the superstitious belief also maintains that if you pass on your clover to someone else, your luck will double. Does it mean if the woman passed her 21 clovers to her friends, her luck multiplied 42 times? That's enough luck to win a million-dollar lottery or, at least, finding a place in the Guinness Book of Records.

Legend has it that Abraham Lincoln carried a four-leaf clover almost every day for good luck. He didn't have his clover on the day he was assassinated. Another legend has it that when during one of his early campaigns Napoleon Bonaparte bent over to pick up a four-clover plant that he has spotted in a field, his head escaped a sniper's bullet.

From superstition to science: The number of leaves in a clover plant is controlled by its genes. The dominant gene characteristics in clover plants is three leaves. The plant also has a recessive gene that creates a four-leaf clover. This gene is masked by three-leaf genes. Plant biologists have now been breading clover plants that produce a relatively large number of four-leaf plants. It seems the 'luck of the Irish' four-clover plant is running out.

58

Murder in odds

Murder is a relatively rare occurrence, and the odds of being murdered are very low compared with other causes of death. Besides, murder statistics depend upon numerous factors, including age, sex, race, culture, location, socioeconomic status, and political situation.

"Murder in odds" seems like a title of Agatha Christie novel. 'The dog hunts rabbits,' writes the so-called Queen of Crime, in *Dumb Witness*. "Hurcule Poirot hunts murderers." If the famous fictional detective digs into the odds of being murdered, he won't find a murderer but numerous victims. The following figures are for 2012 and are from the World Bank's 2012-14 database, which describe murders as "intentional homicides" by individuals or small groups. Killing in armed conflicts is excluded from this database.

Murders per 100,000 people

Australia	1	Germany	0 (1 in 2011)
Brazil	25	India	4
Canada	2	Japan	0 (0 in 2011)
Colombia	31	United Kingdom	0 (1 in 2011)
El Salvador	41	United States	5

You can easily work out which of the 10 countries listed above has the lowest odds of you being murdered. The total number of murders in 2012 was 437,000 worldwide. The odds of you being bumped off work out to be roughly 1 in 16,000.

59

Would you see a supernova in your lifetime?

Astronomers have calculated the odds to be nearly 100% that a supernova would be visible to infrared telescopes sometimes during the next 50 years. Unfortunately, the odds dip to disappointing 20% that this grand stellar spectacle would be visible to the naked eye in the night sky.

A supernova is an old star exploding and brightening near the end of its life. It rises to brilliance very quickly, often in a matter of few days, and then declines in brightness slowly. Supernovas sometimes become bright enough to be seen in the daytime. The remaining matter of supernova forms a neutron star, a star so heavy that even a pinhead of its matter would have mass of millions of metric tons.

In 1054 Chines astronomers noted a new star in the sky which became about four times brighter than Venus in its brightest light. We know now that this "guest star," as the Chinese called it, was a supernova. It was visible in daylight for 23 days.

In 1572, the Danish astronomer Tycho Brahe saw a new star blazing out in the sky. He watched the star – as bright as Jupiter – carefully night after night until it gradually faded. In 1573, he published a small book *De Nova Stella* ("A New Star"") in which he suggested that stars could have a beginning, middle and end. This revolutionary idea smashed the contemporary view that stars were fixed and unchanging. Today we know that Tycho (he is usually known by his first name) observed a supernova that was visible in the northern skies for nearly two years.

It has been suggested that the Star of Bethlehem was a supernova, but most astronomers disagree. Such stellar catastrophes are far too spectacular to not generally notice and, except for Matthew, none of the historical writers of the time mention such a brilliant star near the time Jesus was born. However, according to Chinese and Korean astronomical chronicles, there is a record of the sighting of a bright star in the Far East.

The latest news about supernovas come from astronomers at the Ohio State University. The supernova they have observed is in the Milky Way. Technology has now made it easier for astronomers to study supernovas in our galaxy. Unlike Tycho, they don't have to peer through an ordinary telescope.

Good news for those who live in the southern hemisphere: they have a better chance to see this supernova with the naked eye since more of the Milky Way is visible there.

60

Probability tale of Punxsutawney Phil

The weather forecast of many groundhogs that have been designated beloved Punxsutawney Phil over the years in the town of Punxsutawney, Pennsylvania have been correct 39% of the time.

If you are a fan of the very popular 1993 movie *Groundhog Day*, you know what we're talking about. In the movie, a grumpy television weatherman (played by Bill Murray), who has been reluctantly sent to Punxsutawney to cover a story about a weather forecasting groundhog, finds himself in a seemingly endless time loop living the same day over and over again.

The Punxsutawney townsfolk celebrate 2 February as the Groundhog Day with a festive atmosphere of music and food. On that day, Punxsutawney Phil appears from his temporary home. The tradition has it that if he "sees" his shadow, the town is in for six more weeks of winter; if he doesn't, he has predicted an early spring. The Groundhog Day has been celebrated since 1887.

In their trademark spirit of investigative reporting some nerdy boffins at The Washington Post (the paper is known for digging a hole for President Richard Nixon of the Watergate notoriety) managed to find a little time to dig through 114 forecasts (99 predicting more winter and 15 early spring). They say Punxsutawney Phil's predictions have been correct only 39% of the time in his hometown.

Back to the hapless weatherman, he arrives in the town on 1 February with a female news producer and a male cameraman. When the time loop breaks on 3 February the weatherman finds that he is a different person – and in love with the news producer – than he was on 1 February. The internet movie database IMDB gives the movie a rating of 8.1 (out of 10).

61

Who looks to stars?

There is no reliable global and national data on the odds of people believing in astrology. In the US, around 25% adults believe in horoscopes. In Australia it rises to 40%, while in China, it dips to 8%. About 45% Americans consider astrology "very scientific" or "sort of scientific," according to the National Science Foundation.

Though astrology and astronomy had a starry relationship in the past, astrology is not a science. Obviously, astrologers disagree. They even assign astronomy a lower status as an offspring of astrology. Astronomy has made remarkable progress in the past few centuries, but astrology has hardly changed.

Tell this to astrologers and they would cast the horoscope of their "science": astrology is truth and truth never changes. But we know now that some ancient "truths" are not eternal. For example, when astrologers say that the Sun is in Leo between 23 July and 22 August, they mean that the Sun, as seen from the Earth, is in the same part of the sky as the constellation of Leo.

They ignore the astronomical phenomenon of precession, the slow, conical motion of the Earth's axis of rotation which makes the Earth wobble around its axis in a 25,800-year cycle. Precession makes the Earth's position relative to the constellations of the zodiac change over centuries. The Sun is no longer in the constellation of Leo; it's in the constellation of Cancer. If you were born between 23 July and 22 August, your zodiac sign is no longer Leo; it's Cancer. Shouldn't it affect you, if you are a Leo, or any other astrological sign?

Many studies, however, have shown that being born at certain times of year relates to a small but significantly increased risk of developing certain mood disorders. In a study researchers found that eight of every 100 people born during spring in the northern hemisphere had anorexia compared with 7% those without anorexia. This is a 15% increase in risk for those born during spring in the northern hemisphere.

In another study researchers matched the birth season of 400 people to personality types in later life. They discovered that when you are born may increase or decrease your chance of developing certain mood disorders. For example, people born in the winter were significantly less prone to irritable temperament than those born at other times of the year. The researchers believe that the seasons influence certain neurotransmitters such as dopamine and serotonin which control mood.

"Season affect our mood and behavior," says Eduard Vieta of the European College of Neuropsychopharmacology. "Even the season at our birth may influence our subsequent risk for developing certain medical conditions, including some mental disorders."

These are medical studies trying to find out how seasons affect our mental health; in no way do they support your astrological birth sign.

Why bother reading your stars? They can't change the odds the things that matter in your life.

62

Does your zodiac sign matter when it comes to marriage?

The probability of astrology influencing the success of your marriage is zero. However, astrological signs can affect how people feel about things.

Does astrology work? Many major studies have tried to answer this question, but none has found any proof that the astrological sign you were born under influences the person you become.

The largest test of astrology ever taken – a statistical analysis of the birthdays of more than 20 million married people in England and Wales – has shown that their astrological sign has no impact on the probability of marrying and staying married to someone of any other sign. There is no such thing as zodiac signs and love compatibility. Googling "love sign astrology" would produce hundreds of thousands of hits, but none of them is going to make you wiser in choosing your life partner.

However, a study by British psychologists Susan Blackmore and Marianne Seebold suggests that women's relationships are somewhat affected by what they read in their horoscopes. They write in *Correlation*, a journal of research into astrology published by the Astrological Association, UK.

More generally, the results confirm the strong influence of astrology on women's lives. 72% do not think astrology is just superstition and almost 90% said that they find out the sun signs of people they have relationships with. 78% had read a book concerning their sun sign in love. Even though only 15% said they would alter their behaviour according to what they read in a horoscope, these results suggest that astrology may influence women's behaviour in many ways.

So, Venus, the so-called love planet, moving into your astrological sign might not affect your marriage, but reading those horoscopes at the back of magazines and newspapers, may sway the odds for or against you.

We are suckers for weird beliefs because the brain allows them and even encourages them. Compared to absurd things– alien abduction, ancient astronauts, crystal healing, ESP, magnetic therapy, quantum healing, UFOs and so on – belief in astrology seems like a harmless entertainment.

63

Warts and all

WHIM syndrome is such a rare disease that only 65 cases have been reported worldwide to date. The odds of having this disease are estimated to be 1 in 5 million births. Here's a woman's story of astronomical odds of being cured of this rare disease by fluke.

In WHIM syndrome, the body's immune system doesn't function properly, and patients have defect in a single section of their DNA. Their bodies can still make immune cells, but they get stuck in the bone marrow where they are produced.

More than 80% of the WHIM patients develop, by the age of 30 years old, widespread human papillomavirus (HPV)-induced warts that are often difficult to treat, generally starting on hands and feet. They also they have an increased risk for bacterial infections and a susceptibility genital cancer and liver failure. By the age of 40, the cancer risk is about 30%.

The scientific journal Cell reports of an unidentified woman who has been spontaneously cured in an event so improbable doctors say it is the medical equivalent of a mega-lottery win. The 58-year-old woman was the first identified case of WHIM more than 50 years ago. As WHIM can be passed on to children, the woman sought out a team of researchers at the US National Institute of Allergy and Infectious Disease to get her daughters tested. She had passed the disease down to her two daughters.

But the researchers were shocked to hear when she said that her own warts had disappeared 20 years ago. The researcher then tried to find out how this "miracle" happened. They say "chromosome shattering" spontaneously cured her. In "chromosome shattering" a part of the DNA is rearranged. In her case this led to 164 genes lopped out of her DNA, including the mutated one that was causing the problem. Luckily, "chromosome shattering" took place in a stem cell that makes immune cells. If it had happened in a muscle, it would have made no difference to her condition.

Researchers hope a better understanding of the disease could lead to a treatment or even better ways of carrying out bone marrow transplants.

7.8 billion and counting

A new population prognosis: the world population will continue growing this century. There is 80% probability that world population will increase to between 9.6 billion and 12.3 billion in 2100.

Population projects are based mostly on two things: future life expectancy and fertility rates. The current projections are about 2 billion higher than widely cited previous estimates, which say that the world population would go up to 9 billion and level off or probably decline. The new projections are based on modern statistical tools developed by the University of Washington, which uses data collected by the United Nations.

The most increase in population is expected in Africa, where there is 95% chance that population at the end of the century will be between 3.5 and 5.1 billion people. Asia, now 4.4 billion, is projected to peak around 5 billion people in 2050 and then begin to decline. Populations in North America, Europe and Latin America and the Caribbean are projected to stay below 1 billion each.

The world population is now nearly 7.8 billion (you can see continuously updated real-time population data, including births and deaths today at *www.worldmeters.info*, which claims that its data comes from "most reputable organizations and statistical offices in the world").

Pi face

What are the odd of someone calculating the value of π to 527 correct digits (a pencil-and-paper calculation without the help of a computer)? The answer is 1 in 107.7 billion. Only one person in history has done it and demographers estimate that about 107.7 billion people have been born since 8000 BC.

All circles are similar and the ratio of the circumference to the diameter is always the same number. This ratio is known as π (the Greek letter pi). It takes infinite digits to express it as a decimal number. It is impossible to find the exact value of pi; however, the value can be calculated to a very high degree of accuracy.

In 1650, the English mathematician John Wallis worked out unlimited series for the calculation of the value of π. This opened a new crazy field in mathematics – calculation of the value of π to an unlimited decimal place. The craze continues, but the work is now done by supercomputers.

In 1853 William Shanks, an English mathematician, published a small book containing a value of π that began with the familiar digits 3.14159 and went on for 707 places. He worked on his calculations for nearly 20 years. Computer analysis now shows that Shanks` calculation has errors that begin in the 528th decimal place. The first 527 digits are absolutely correct, which were not bettered for almost a century. The record so far is calculating value to over 13.3 trillion digits (by a supercomputer, not a superhuman).

Why does one compute π to trillions of digits, when even for designing a space probe one need not know the value to more than a few digits? Like Edmund Hillary, of Mount Everest fame, any π fan would say, "Because it's there?"

In 1777, the French naturalist and mathematician Georges-Louis Leclerc, Comte de Buffon tried a novel experiment to determine the value of pi. The experiment – the first experiment in geometrical probability – involved ruling equidistant parallel lines on a plane horizontal surface. He then he dropped a needle with a length equal to the distance between the lines repeatedly on the ruled surface. If the needle crossed or touched a line, the toss was considered favorable.

Buffon said that the probability of the needle crossing or touching the line is $2/\pi$. If the needle is tossed at random n times and it crosses or touches c times, then $2n/c$ will approach π, if you keep on dropping the needle. The more drops, the more closely the result will approximate π.

In 1901 Mario Lazzarini, an Italian mathematician, tossed a needle randomly 3408 times and observed 1808 hits. From these figures, he arrived at a value of 3.1415929 for π, which is correct to six decimal places. Subsequent similar experiments by other investigators have resulted in less accurate values of π. Some mathematicians now suspect that Lazzarini faked his results! Remember the mnemonic "how I wish I could enumerate" and you wouldn't be faking to recall the value of π to five decimal places.

66

Too many choices spoil the jackpot

Given too many choices, you are more likely to overestimate your chances of winning a jackpot.

You may think that having more choices is always a good thing. But a recent study proves otherwise: given too many choices, you're more likely to make a bad, risky decision. More choices mean more information, but information load caused by too many choices results in information fatigue and interrupts our ability to make well-informed and balanced decisions.

As gamblers make most bad decisions, psychologist Thomas Hills of the University of Warwick in the UK and his colleagues set up a decision-making test based on gambling. Each person in a group of 64 university students clicked on a box out of several shown on a computer screen. Each box could pay out an amount ranging from £1 to £5 with certain odds of paying out.

The participants were allowed to sample the boxes as many times they wished and decided which box to choose to win the jackpot. Clicking on a box sampled showed the amount of payout and the odds for that payout. The participants were assigned to either "many-to-few" or "few-to-many" conditions. One group started choosing between 2 boxes and the number went up to 4, 8, 16 and 32. For the other group the number decreased from 32 to

16, 8, 4 and 2. In both groups the participants took more samples with larger set sizes. But when they were presented with that many choices, they didn't study the odds carefully enough to get a better idea of the payout odds. They ended up betting on the high-risk, high-payout box.

The researchers say that despite consistent understanding of information about odds, changes in searching for that information had dramatic effect on risky choice as the number of boxes increased, with participants tending to choose gambles with rarer chances of winning.

Ockham's razor – a rule developed by William of Ockham who lived in the thirteenth century in England – can help us thinking clearly when we are faced with too many choices. The rule – it is vain to do with more what can be done with less – implies keep the number of causes and explanations in your decision-making to a minimum. Advice to computer programmers to keep their programs simple – keep it simple, stupid – is in similar vein. Sometimes it's better not to devour too much information. Too many choices can make you miserable.

Whoever said "it's choice, not chance, that determines your future" was probably right.

67

Another sunny day, another odds-on lucky day for skin cancer

One in every three cancers diagnosed is a skin cancer. At least two in three Australians will be diagnosed with skin cancer by the time they are 70. It's one of the highest incidences of skin cancer in the world, two to three times the rates in Canada, the US, and the UK. In Japan and China incidences of skin are much lower.

Anyone can be at risk of skin cancer, though the risk increases as you get older. Most skin cancers are caused by exposure to ultraviolet (UV) radiation in the sun. Melanoma, the most-deadly form of skin cancer, is associated with sunburn. Sunburn is a sign of short-term overexposure to UV radiation. You can also be sunburnt on cooler or overcast days when you would mistakenly believe UV radiation in not as strong.

UV rays are strongest in areas closer to the equator. Because the sun is directly over the equator, UV rays only travel a short distance through the atmosphere to reach these areas. The ozone layer, which absorbs the sun's harmful radiation, is naturally thinner near the equator. Australia's relative proximity to the equator, makes it people overexposed to UV radiation.

Everyone, regardless of skin color, can get sunburn. Some individual risk factors for skin cancer, according to the World Health Organization, are:

- fair skin
- blue, green or hazel eyes
- light-colored hair
- tendency to burn rather than suntan
- history of severe sunburns
- many moles
- freckles
- a family history of skin cancer

"Slip Slop Slap" slogan launched by the Australian Cancer Council in the 1980s to prevent sun damage is still relevant when it comes to avoiding skin cancer. The slogan was updated in 2007 to "Slip Slop Slap Seek Slide" (slip on a shirt, slop on sunscreen, slap on a hat, seek shade, slide on wraparound sunglasses).

68

The sky is not falling

The odds of a single person severely injured by a re-entering space debris from a satellite in a given year are about 1 in 100 billion, according to the European Space Agency. In the course of a 75-year lifetime the odds are a little less than 1 in 1 billion.

Since the launch of Sputnik 1 in 1957 some 15,000 metric tons of man-made space objects have re-entered the Earth's atmosphere without causing a single human injury. In 1997, an American woman became the first – and still the only – person hit by a piece of space junk. A DVD-size piece of metal struck her shoulder. Because of wind resistance it came down so slowly that she wasn't hurt.

Millions of pieces of debris – from inactive satellites, launch vehicles and other discarded parts left over from separation – remain in orbit high above the Earth's atmosphere. NASA has tracked more than 500,000 pieces of space junk as they orbit the Earth. They all travel at about 28,000 km/h (17,500 mph), fast enough for a small piece to damage a satellite or spacecraft. Even tiny paint flecks can damage a space craft when travelling at such a high speed.

In 2013, a satellite weighing about 1,100 kg (2,425 lb) and about 5.3 m (17 ft) long disintegrated and burned in the atmosphere. The remaining fragments – about 25% of its total mass – were scattered over the Southern Atlantic Ocean near the Falkland Islands. An Islander managed to catch it on a camera as it disintegrated. This European Space Agency satellite, which was at the end of its natural life when it ran out of fuel, was one of the biggest satellites to re-enter the Earth.

69

If you're not dreaming, you're daydreaming

There is 1 in 10 chance that you're dreaming – or 1 in 4 chance that you're daydreaming – right now.

Most dreams, but not all, occur during the rapid eye movement (REM) period of sleep. We spend nearly one-thirds of our lives asleep, and one-fifth of this time is REM sleep. This means we spend one-fifteenth – that's a big slice – of our lives dreaming.

This works out to 1.6 hours of dreaming every night. If you're awake 16 hours a day, the chances of you dreaming at any given moment are 10%.

REM sleep is the fifth and the last stage in five stages of sleep. It lasts 10 to 15 minutes; accompanied by rapid, jerky eye movements; heart rate, blood pressure and body temperature become much more variable; the brain is highly active (it's on fire); vivid dreams occur (dreams also occur in other stages but they are not vivid).

During the night, these stages of slow-wave and REM sleep are repeated in roughly 90-minutes cycles until waking occurs. There is more-slow wave sleep early on and more REM sleep towards morning.

The other four stages of sleep are:

Awake: Sleep-on neurons, a small group of neurons in the

forebrain responsible for inducing sleep, are inactive; alpha brainwaves (relaxation)

Stage 1: Marks the transition between awake and asleep; sleep-on neurons fire; shallow brainwaves; muscles relax, and eye movement slows down

Stage 2: It lasts the longest; bursts of wave activity; sleep talking

Stages 3 and 4: It last for about 30 minutes; deepest, or slow-wave, sleep; delta waves appear; sleepwalking and bedwetting.

Like dreams during sleep, daydreams play an important role in our lives in during our waking hours.

We daydream for about one-third of our waking hours, although a single daydream lasts only a few minutes. This means we are daydreaming for 16/3 or 4 of our waking hours. Your chances of daydreaming at any given moment are 25%.

The human brain has been purposefully designed – or hardwired, as they say these days – for daydreaming; it is our minds' default mode of thought. Brain scans show that our brains have a "default network," a region which remains active when we are daydreaming.

When we have a specific task, our minds focus on that task. But most of the time we are engaged in less directed, unintended thoughts and that state is routinely interrupted by periods of goal-directed thoughts. The human brain prefers its default mode. But it immediately springs into action when some task requires attention.

70

Your digital self

In the digital world, the odds are stacked against your uniqueness. How much information does it take to single out one person among billions? The answer is: just 33 bits of information.

Self is always obvious to us, but to others it's our identity that makes us different from them. Each of us has a biological identity which is evident in our DNA sequence and biometrics such fingerprints, facial structure, the iris, and voice. We also have a legal identity (name, date and place of birth, signature and so on). Now a vast majority of people on the planet also have a digital identity, the data that uniquely describes a person.

You may be aware of your identity on Facebook, Twitter, and other social media sites, but you probably do not know of many of your other digital fingerprints. Did you know that the history of the websites you have visited is also a unique identity? Even if you delete your browser history regularly, it leaves it paw marks somewhere in the cyberspace, for an eternity as far as we know.

To check the uniqueness of my browser fingerprint I logged into the website *panopticlick.eff.org* and came back with the answer "your browser fingerprint appears to be unique among the 5,021,930 tested so far." It also advised me that my browser fingerprint conveys at least 22.26 bits of identifying information (a bit is the smallest unit of information, it stores just a 0 or 1; eight bits make one byte).

Just 33 bits of information is enough to single out any person from the world population of 7.3 billion, says American science writer Brian Hayes in *American Scientist* magazine. "In a world where every tiny idiosyncrasy can be catalogued and filed away in milliseconds, it's all too easy to compile a unique fingerprint."

According to this calculation, I'm less than 11 bits away before I lose my "right" to my digital identity. Anyone with those 33 bits of information can not only identify me but also masquerade as me.

Some governments are working on regulations to control "device fingerprinting," a process of silently collecting information about the users of computers, tablets, and smartphones. Each device connected to the internet identifies itself in various ways to help websites to deliver the required information. This information in conjunction with other data such as the internet connection can identify a user.

To protect personal information, the Australian Cybercrime Online Reporting Network offers the following advice.

- Activate privacy settings in social networking sites.
- Never reveal details that might identify you such as full name, date of birth, place of birth, address or contact numbers.
- Don't post anything you don't want strangers to know or find out about.
- Think before you post online as it can never completely be deleted.

71

Twin fingerprints

What are the odds of identical twins having identical fingerprints?

Identical twins have the same DNA and thus the matching physical features, but not the same fingerprints. Every person has a unique set of fingerprints making them the ideal way to identify a person, even identical twins. So far no one has come across any two identical digits.

Out fingerprints are set for life around the 19th week of growth in the womb, roughly halfway through a normal pregnancy. The fine details of our fingerprints – whorls, arches, ridges, and valleys – are determined both by genes and the local environment around the dividing skin cells. Twins grow in the same uterus, but they have different lengths and diameters of umbilical cord. Even a slight change in a twin's individual umbilical cord will cause slight difference in fingerprints.

Researchers have found that differences in fingerprints between the thumb and little finger are associated with likelihood of developing diabetes later in life.

Why do we have fingerprints? Researchers have debunked the popular notion that fingerprints help us grip more firmly. They reduce friction between skin and surfaces by reducing the area in contact with surface by about one third. Fingerprints also improve the hand's tactile sensitivity. They also help wick water off our hands, improving grip on wet surfaces. Of course, they help in solving crimes.

Only a handful of people in the world do not have fingerprints. This is caused by a rare genetic mutation.

72

A surprising law of digits (as in numbers)

If your house number starts with 8, about 5% house numbers have 8 as their first digit.

The same holds true for any set of numbers you find anywhere and have nothing in common: populations, sports statistics, stock market prices, death rates, length of rivers, answers in physics book, size of files stored on your computer, sales figures, prime numbers, and so on. It's true for any group of non-random numbers.

The first or "leading" digit of each number is not random and appears with equal frequency. This surprising trend in number sets was discovered it in 1938 by physicist Frank Benford while working at the General Electric Research Laboratory in New York.

First digit	Frequency
1	30.1%
2	17.6%
3	12.5%
4	9.7%
5	7.9%
6	6.7%

7	5.8%
8	5.1%
9	4.6%

For years Benford's law was seen as a mathematical curiosity, but it's now used to analyze all sorts of data to detect fraud. It can reveal suspicious data in drug trials to electronic voting. Fraudsters usually fiddle with just a part of data set, but they don't know how the data will be analyzed by Benford's remarkable law.

73

A surprising law of digits (as in fingers)

The 'power of prediction' of your finger ratio: the length of the index finger divided by the length of the ring finger says a lot about you.

Take a ruler, stretch your right-hand palm up and measure the length of index and ring fingers, starting from the crease nearest your palm to the tip of the finger. Or, you can take measurements from a photocopy of the hand. Now divide the length of your index finger (second digit, 2D) by the length of your ring finger (fourth digit, 4D). The result is known as second-to-fourth-digit-length ratio (2D:4D), or simply the finger ratio. In most people both hands have slightly different finger ratios. The average ratio for women is above 1 and for men less than 1.

In the early stages of pregnancy, the womb is washed over by sex hormones estrogen and testosterone. If you were exposed to more estrogen than testosterone in the womb, the index finger will be longer than the ring finger. If you were exposed to more testosterone than the ring finger will be longer.

Our early exposure to estrogen and testosterone influences how our body develops. There is growing evidence that it may affect predisposition in later life to disease and sexual orientation. Scientists have discovered a link between prostate cancer and finger ratio: men with higher ratios run a significantly higher risk of prostate cancer. Women with lower ratios are more susceptible to osteoarthritis and cervical cancer. Gay men seem to have significantly higher ratios than heterosexual men, and lesbian have a low ratio. This indicator is not strong enough to allow you to jump to conclusion about someone's sexual orientation by just looking at their fingers.

According to American anthropologist Helen Fisher, your finger ratio says something about your personality. If you have a longer index finger, "you have good verbal skills, can find the right word rapidly, are good at remembering, better at compassion, nurturing, patience, have good people skills." If you've a longer ring finger, "you tend to have poorer social skills but be direct, decisive, ambitions, competitive." Both men and women with a low finger ratio tend to be assertive.

Korean researchers have stretched the "power of prediction" of finger ratio a bit too far by linking it to penis length. They have found that men with a lower ratio tended to have a longer penile length. Their result is based on the study of 144 men twenty or older who were hospitalized for urological surgery. Their results: the average flaccid and stretched penile lengths were 7.7 cm (3 inch) and 11.7 cm (4.6 inch) respectively, while the average finger ratio was 0.97. Don't rush to generalize these results. The Korean study was conducted on a single ethnic group of men, and digit ratio has been shown to vary among ethnic groups.

74

Don't confuse a game of chance with a game of skill

National lotteries and casino games operate according to laws of probability, but horse racing and other sports involve skills of horses or humans which defy prediction. Determining odds in a game of skill is an art, not science.

If you are a fan of football or any other sport for that matter, you would think that you have a better chance of winning a bet because you have an in-depth knowledge of the game. You are wrong.

Researchers at Tel Aviv University designed a study to test how much specialized knowledge benefits gamblers. Participants were divided into three groups: 53 sports gamblers with specialized knowledge base having answered 80 to 100% of questions correctly about football; 34 amateurs who loved football and answered 60 to 80% questions; and 78 laypersons who didn't gamble and who answered less than 60% of the questions. They all were asked to predict in advance the general result and the exact result of football matches in the European Championship League Round.

The study result showed that people who had prior knowledge of football above and beyond "the aim is to kick the ball into the goal" fared worst. Instead, the researchers found "no significant difference between the groups." Pathological sports gambler who seems to have an "illusion of control" attained by knowledge of the game and its latest data are simply victims of their "magical thinking."

Determining the right odds of winning in horse racing is an art, not science. "There are people who try to make a living betting on horse racing," says Erik Snowberg, a professor of economics at the California Institute of Technology who has studied horse racing. "There are people who genuinely have better information and can utilize that information to make a profit. You can think of it as the same way as people who have models to predict the stock market. People have models to predict the outcomes of horse races."

The question is how to get your hands on genuine information. This genuine information might be accurate for horses to some extent, but it's nearly impossible to predict the behavior of individual humans in a team game. Besides genuine information good gamblers have genuine control on their gambling: they know when not to bet, they evaluate their winning odds every time.

Don't place your bet on it but research shows that horses are right or left "handed." A study of 40 unschooled sport horses found that majority of female horses seemed to prefer their right side, while the majority of male seemed to prefer their right side. A very small number of the horses used both sides. Would learning horse handedness would make you a winner at horse races? We simply don't know.

75

Authors' odds of success

Publishing seems like a catch-22 situation: there is no way to work out odds of getting published while publishing a book is so much dependent upon chance.

News snippets such as *To Kill a Mockingbird* has sold more than 40 million copies worldwide or *Fifty Shades of Grey* trilogy more than 100 million copies make people curious about an author's odds of success.

Since 1455 when Johannes Gutenberg printed the first book – the Bible in Latin – on his hand-printing press nearly 130 million different books have been published worldwide. The number of people born since the dawn of history is estimated to be about 107.7 billion. This means about 1 in 828 people who have ever lived on this planet or living now is an author. Don't hasten to conclude that the odds of getting published are 1 in 828. Odds don't work that way.

Look at these figures from the International Publishers Association.

Country	Number of books (new titles + re-editions) published in 2013	New titles per million inhabitants
China	444,000	325

United States	304,912	959
United Kingdom	184,000	2875
Russia	101,981	699
Japan	77,910	613
France	66,527	1008
Indonesia	30,000	119
Australia	28,234	1176
Brazil	21,085	104
Pakistan	3,500	19

It would seem that the odds favor authors in the UK as publishers over there churn out more book per capita than any other country. Certainly, it might be easier to get a book published in the UK, but that rule doesn't apply to everyone. From the above figure the odds of getting published in the UK works out to be around 1 in 348, but publishing is not a lottery, not even a competitive sport. There are no odds. Tell this to Americans: a poll shows that 80% of them would like to be an author. I assure you this figure doesn't include newly born children who cry out for a book contract before they do for milk.

Publishing seems like a catch-22 situation: there is no way to work out odds of getting published while publishing a book is so much dependent upon chance. You'll understand what I mean if you read the following rejection counts of some mega-sellers: Jack Canfield and Mark Victor Hansen's *Chicken Soup for the Soul*, 140 times; Margaret Mitchell's *Gone with Wind*, 30 times; Stephen King's *Carrie* 30 times; James Joyce's *The Dubliners*, 22 times; William Golding's *The Lord of Flies*, 15 times; Stephenie Meyer's

Twilight, 14 times; J. K. Rowling's *Harry Potter and the Philosopher's Stone*, 12 times; L. M. Montgomery's *Anne of Green Gables*, 5 times ... the list goes on but one can only gasp at the words of one of the 15 publishers who rejected *The Diary of Anne Frank*: "The girl doesn't, it seems to me, have a special perception of feeling which would lift that book above the 'curiosity' level."

76

Miracles happen every day

You can expect to experience an event with odds of 1 in 1 million at the rate of about once per month during your lifetime.

Back in 1955 John Littlewood, a Cambridge University mathematician, reasoned that a miracle is a one-in-a-million lifetime event.

During the time we are awake and actively engaged in living our lives, about eight hours each day, we see and hear things happening at a rate of about one per second. So, the total number of events that happen to us are about 30,000 per day or about a million per month.

The law of large numbers says that with a truly large sample, any "miraculous" thing is likely to happen. If we take the world population as nearly 7.8 billion, then we expect 7.8 billion such occurrences a month or 213 million miracles a day. Accordingly, seemingly miraculous events are commonplace.

Littlewood enjoyed debunking mysterious ideas. In his book, *A Mathematician's Miscellany*, first published in 1953, he writes:

Certain ancient Indian writings reveal an awestruck obsession with immense stretches of time ... There is a stone, a cubic mile in size, a million times harder than diamond. Every million years a very holy man visits it to give it the lightest possible touch. The stone is in the end worn away. This works out at something like 1035 [1 followed by 35 zeroes] years; poor value for so much trouble, and an instance of the "debunking" of popular immensities.

77

We may like to win, but we hate to lose

Why are people unwilling to accept a bet with odds of 1 to 1 unless the amount they could win is nearly the twice the amount they might lose?

Years ago, psychologists Daniel Kahneman and Amos Tversky showed why people are reluctant to bet on a fair coin for equal stakes. They called it loss aversion: the attractiveness of the possible gain is not nearly sufficient to compensate for the aversiveness of the possible loss. In our decision making, losses loom larger than gains.

In an in interview with *The New York Times*, Kahneman (now a Nobel Prize winner but better known for his highly popular book Thinking, Fast and Slow); "In my classes, I say: I'm going to toss a coin, and if it's tails, you lose $10. How much would you have to gain on winning for this gamble be acceptable to you?

"People want more than $20 before it is acceptable. And now I've been doing the same thing with executives or very rich people, asking about tossing a coin and losing $10,000 if it's tails. And they want $20,000 before they will take the gamble."

Like confirmation bias – seeking supporting evidence but ignoring discomforting evidence – loss aversion is one of the most potent of our cognitive biases. These biases control our behavior and emotions which in turn color our decisions. Discounting by stores is a surreptitious way to go around our loss aversion bias. Sale prices exhort that the low price won't last forever, and our unconscious mind focuses on the fear of losing the deal.

78

Prediction is impossible

Chaos theory provides a bridge between the laws of physics and the laws of chance. Can it predict the throw of a dice?

Jules-Henri Poincaré was in the habit of buying fresh bread every day from his local baker in Paris. He suspected that the bread weighed less than the advertised weight of one kilogram. He started weighing the bread daily at his home. After a year, he plotted the graph of daily weights, which showed a bell-curve with the minimum weight of 950 grams but truncated on the left side of the kilogram mark. He reported the matter to the authorities.

The anecdote, probably apocryphal, gives an insight into the renowned French physicist and mathematician's life-long quest for beautiful mathematical patterns. He said that 'it may be very hard to define mathematical beauty, but that is just as true of beauty of all kinds.'

Years later, in 1908, Poincaré made a remarkable observation which led to the foundation of the new science of chaos: "Small differences in the initial conditions produce very great ones in the final phenomena. A small error in the former will produce an enormous error in the latter. Prediction becomes impossible."

Poincaré's observation received little attention from his contemporaries but has now earned him the title of the "founder of chaos theory" as we know now that the behavior of a dynamic system depends on its small initial conditions.

Chaos describes disorderly systems. The behavior of a chaotic system is difficult to predict because there so many variable or unknown factors in the system. Chaos is a dynamic phenomenon. It occurs when the state of a system changes with time. Even simple systems can grow exponentially with time, making long-term prediction of the future impossible. The behavior of a dynamic system depends on its small initial conditions. In a chaotic system, even a small change can bring about major upheaval. Chaos helps scientists to understand the complexities of nature as it provides a bridge between the laws of physics and the laws of chance.

The wider significance of Poincaré's observation was recognized in 1963 when Edward Lorenz, a meteorologist, developed a computer model to predict weather patterns. His model suggested that even a small initial unpredictable condition such as a flapping butterfly could produce a larger global change in weather.

This is now called the "butterfly effect": an action as small as a butterfly flapping its wings, say in Beijing, could bring about a snowstorm weeks later thousands of kilometers away in New York. The butterfly effect may be somewhat fanciful exaggeration of chaos, but when we combine uncertainty with complexity the results are totally unpredictable. If you have the romantic notion that the butterfly effect can be applied to predict the outcome in a gambling situation, you're wrong. The accuracy with which the initial conditions (say speed and momentum of a roulette ball and the wheel it is rolling around in it) would have to be determined to predict the outcome (where the ball will land) is beyond any gambler's capacity.

Chaotic behavior occurs in phenomena as diverse as the stock market, disease epidemic, population changes and the human heartbeat. Chaos theory, which touches all disciplines of science, can be used to examine the apparently random unpredictable features of the everyday world, such as the turbulent flow of water, traffic jams, the path of the winds and the build-up of clouds.

Researchers from the Technical University of Lodz in Poland believe they can predict the throw of a dice: where the dice will land, and which side is facing up. They suggest that the toss of a symmetrical dice is not a perfectly random event, but the accuracy required for determining the initial position is so difficult and "exact" that the toss is still effectively random. You have to be magician to throw a dice in the way to get the desired result.

It's better to apply chaos theory to create ripples of happiness than applying it to a casino game. Some psychologists suggest that the basic idea of chaos also works in our personal lives: a small act of kindness causes a small ripple and if there are enough small acts, they could magnify into the butterfly effect of happiness making people happier on the other side of the street, the town, the world.

79

Maximize odds on your investment

**Take the following advice on improving odds of better
return with a grain of salt.**

"Jesus saves but he couldn't on my wage," laments a graffiti artist
on a wall. Remember the first maxim of managing your money:
Money grows only when it is saved and invested wisely.

All money matters have disclaimers in fine print. Here's our
fine print: Although maxims and rules of thumb encapsulate
knowledge, they are not universal truths that can be applied to
every investment situation.

The rule of 72. The rule of 72 says that to find the number of
years required to double your money at a given compound interest
rate, you can just divide 72 by the annual interest rate. For
example, to find out how long would it take to double your money
at 6 per cent interest, divide 72 by 6 and you get 12 years. The rule
also works backwards. You can use it to calculate the interest rate if
you know your money would double in so-and-so many years by
dividing 72 by the number of years.

Don't put all your eggs in one basket. You can spread the risk by putting your money to work in different investment types – shares, managed funds, property, bonds, and collectibles such as art and antiques. You can reduce the risk further by widely spreading within each investment type. For example, you can diversify investment in shares by investing in national shares, international shares, industrial shares, resources shares, large "blue chip" company shares and small company shares. Also, spread your risk over time; that is, spreads your buying and selling over a longer period.

Past performance is not an indicator of future performance. Most financial institutions advise that caution should be exercised in relying upon past performance as an indicator of future performance.

Don't follow a particular investment or friend like a sheep. Make your investment decisions based on your circumstances and advice from investment professionals.

When am I ready for investments? This old rule-of-thumb can help you decide when the right time to make some investments is: When you have saved three to six months' income, you can take some risk with the additional money you save.

The rocket science of loan payment. In the 1970s "rocket scientists" (physicists and engineers who began to work in the financial markets after the downscaling of NASA's space program) applied complex mathematical modelling to understand how finance markets work and how the economy behaves. One of their models show that when paying off a mortgage loan try to pay a little bit extra each month instead of the minimum monthly amount.

"100 minus age" rule. This popular rule-of-thumb concerns the type of investment you should consider: Subtract your age from 100, and that is the percentage you should invest in high-risk growth investments such as shares. The balance should be put in low- to moderate-risk investments such as fixed-interest bonds. If you are about 50, investments should be spread evenly: one-third in high risk, one-third in moderate risk and one-third in low-risk investments. Many financial planners do not like this rule as it leaves no room for the individual choices, longevity, or inflation. They also advise because of increase in longevity we should subtract age from 110. (*See table next page*)

Age	Subtract from 100	High risk	Low risk
30	70	70%	30%
45	55	55%	45%
55	45	45%	55%
65	35	35%	65%

80

Sssh! Wanna buy some happiness?

Once basic needs are met, additional money doesn't make people happier. Or does it? What are the odds of a higher income – or more friends – making you happier?

In the Western culture, personal happiness is one of the most important values in life. In contrast, in some cultures such as in Japan individual pursuit of happiness is perceived as being selfish because it's considered at odd with the good of society.

The elements that play limited roles in our happiness are: money (once basic needs are met, additional money doesn't upgrade happiness), a good education doesn't necessarily lead to happiness, marriage presents a mixed picture, and youth has no advantages over old age.

You might have heard of the world happiness index and there are many ways of calculating it. A recent Gallup survey of 1,000 people in 138 countries, aged and over 15 and over, measured whether they smiled or laughed a lot, learned or did or something interesting, felt respected or experienced enjoyment the day before the survey.

Paraguay topped the list with a score of 87 out of 100. Canada, New Zealand, and Australia each scored 79, the US 78, China 76, the UK 73 and Japan 71. Syria was at the bottom of the list with a score of 36. Nine of the happiest countries are from Latin America because of their focus on the positives of life. The global positive experience index was 71.

It's difficult to quantify happiness in terms of income, but a US study has found a higher income can improve a person's overall happiness, but only up to $75,000. Above this amount – which would undoubtedly vary widely from country to country and region to region – makes no difference.

Another US study points out that an additional $5000 income bumps up the chances of personal happiness by only 2%. But a happy friend of a friend of a friend increases the chance of happiness by 6%. Yet another study – yes, you guessed it right, it's also from the US – has found that doubling your group of friends has the same effect on your wellbeing as a 50% increase in income. Instead of asking your boss for a raise, ask her or him to become your friend.

The message is straightforward: Once basic needs are met; additional money doesn't make people happier. The odds of a super high income making you super happy are indeed 1 in a supremely high number. Not a bet worth making.

The element that genuinely lifts the spirit is strong ties to friends and family and commitment to spending time with them. If you still need convincing about the importance of friends in your life, the results a study of published in the *British Medical Journal* will leave you without any doubt in your mind.

- Knowing someone who is happy makes you 15.3% more likely to be happy yourself.

- A friend of a friend increases your chances of happiness by 9.8% and even your neighbor's sister's friend can give you a 5.6% boost.

The study questioned 12,067 people, who were linked to each other through 53,228 social ties, three times over a period of 20 years.

Another British study published in the online Journal of Epidemiology and Community Health culled its data from a longitudinal study that has followed a group of people born in the same week in 1958, starting at age 7. The researchers examined the response of 6,500 Britons at age 42, 45 and 50. They found that 40% of men and about 30% women had more than six friends who they saw regularly. These men and women were significantly happier than their peers.

"Every form of happiness is private," says novelist Ayn Rand. It's true but a large chunk of our personal happiness that seeps through our lives comes from, not the amount of money, but the number of friends, we have. This fact is neither odd nor measurable in terms of odds.

Let's roll a dice to learn the mathematics of chance

We can find the probability of an event by dividing the number of ways in which the event can happen by the total number of possible outcomes.

$$\frac{\text{number of choices}}{\text{total number of possible outcomes}}$$

This rule can be applied to tossing coins, rolling dice, dealing cards, or drawing lottery numbers. Let's take an example: What is the probability of drawing an ace of hearts from a well-shuffled pack of cards? There are four aces in a pack of 52 playing cards. The probability of drawing an ace is $4/52$ or $1/13$. The probability of drawing an ace of hearts is $1/52$.

A dice has six faces, numbered 1, 2, 3, 4, 5 and 6. The probability that any one of these numbers comes up is $1/6$.

What would be the probability of getting either a 3 or a 5? Because 3 and 5 cannot occur together, such an event is called a mutually exclusive event. In such events, probability is calculated by adding individual probabilities. The probability of getting either a 3 or a 5 is $1/6 + 1/6 = 1/3$.

When two dice are rolled separately, the second dice does not consider what the first dice has done to decide what it will do. Such an event is called an independent event. In independent events, probability is calculated by multiplying independent probabilities. Therefore, when two dice are rolled separately, the odds of getting a double 6 are $1/6 \times 1/6 = 1/36$.

The complement of an event is its opposite; for example, if in tossing a coin "the coin shows heads" is the event, then the complement is "the coin shows tails." The probability of the complement of an event equals 1 minus the probability of the event. In other words, the probability of getting no double is $1 - 1/36 = 35/36$.

After this one-minute lesson in probability, we're ready to look at Chevalier de Méré's problem described in the introduction to this book.

Single dice
Probability of a 6 = $1/6$
Probability of a number other than 6 = $5/6$
Probability of no 6 in four rolls = $5/6 \times 5/6 \times 5/6 \times 5/6 = 0.48$
Probability of at least one 6 in four rolls = $1 - 0.48 = 0.52$ or 52%

De Méré's chances of winning his bet were 52%. The odds, as punters say, were in his favor.

Two dice
Probability of a double 6 = $1/36$
Probability of no double 6s = $35/36$
Probability of no double 6s in 24 rolls = $(35/36) \times (35/36) \times \ldots$ multiply 24 times = 0.51
Therefore, Probability of at least one double 6 in 24 rolls = $1 - 0.51 = 0.49$ or 49%

De Méré's chances of winning his bet were 49%. The odds were against him.

Two dice odds

Number	Combinations (see diagram next page)	Probability (as percentage)	Odds	Bookmakers' odds
12	1	2.78%	1/36	35/1*
11	2	5.56%	2/36 or 1/18	17/1
10	3	8.33%	3/36 or 1/12	11/1
9	4	11.11%	4/36 or 1/9	8/1
8	5	13.89%	5/36	31/5
7	6	16.67%	6/36 or 1/6	5/1
6	5	13.89%	5/36	31/5
5	4	11.11%	4/36 or 1/9	8/1
4	3	8.33%	3/36 or 1/12	11/1
3	2	5.56%	2/36 1/18	17/1
2	1	2.78%	1/36	35/1

* There are 35 chances of losing and only 1 chance of winning, which means chances of winning are 1 in 36 (probability 2.78%)

Two dice combinations

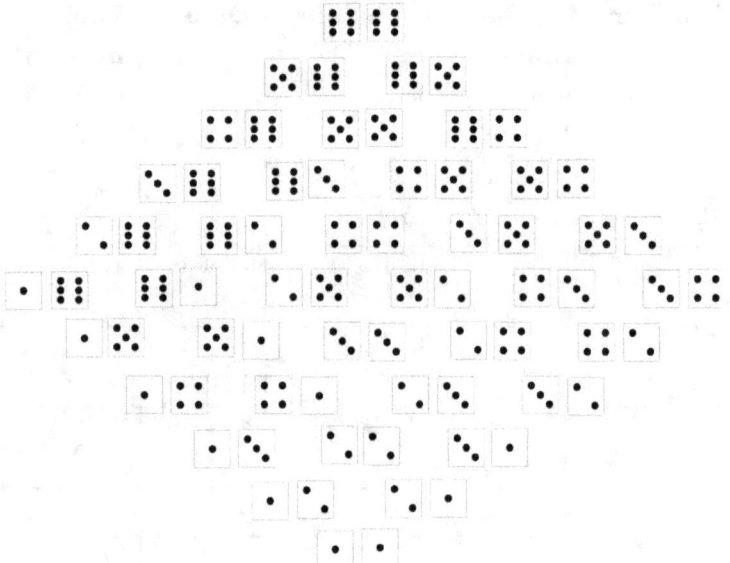

About the author

Surendra Verma is a science writer, journalist and author based since 970 in Melbourne, Australia. He has published numerous popular science books internationally which have been translated into 14 languages other than English. His recent books include:

The Mystery of the Tunguska Fireball
Why Aren't They Here: The Question of Life on Other Worlds
The Cause of Mosquitoes' Sorrow: Beginnings, Blunders and Breakthroughs in Science
The Little Book of Scientific Principles, Theories & Things
The Little Book of Maths Theorems, Theories & Things
The Little Book of Unscientific Propositions, Theories & Things
The Little Book of the Mind: How We Think and Why We Think
Learn & Unlearn: The novel way to rethink the things that matter in your life
Science in 100 Words

+ a children's book
Who Killed T. Rex?: Uncover the mystery of the vanished dinosaurs